T0192119

A Guide to Online Pharmacy Education

A Guide to Online Pharmacy Education

Teaching Strategies and Assessment Methods

Yaser Al-Worafi

CRC Press
Taylor & Francis Group
Boca Raton London New York

CRC Press is an imprint of the
Taylor & Francis Group, an **informa** business

First edition published 2023
by CRC Press
6000 Broken Sound Parkway NW, Suite 300, Boca Raton, FL 33487–2742

and by CRC Press
4 Park Square, Milton Park, Abingdon, Oxon, OX14 4RN

CRC Press is an imprint of Taylor & Francis Group, LLC

© 2023 Yaser Al-Worafi

Library of Congress Cataloging-in-Publication Data
Names: Al-Worafi, Yaser, author.
Title: A guide to online pharmacy education: teaching strategies and assessment methods/ Yaser Al-Worafi.
Description: First edition. | Boca Raton: CRC Press, 2022. | Includes bibliographical references and index.
Identifiers: LCCN 2021061421 (print) | LCCN 2021061422 (ebook) | ISBN 9781032136882 (paperback) | ISBN 9781032136929 (hardback) | ISBN 9781003230458 (ebook)
Subjects: LCSH: Pharmacy—Study and teaching. | Computer-assisted instruction. | Pharmacology—Data Processing.
Classification: LCC RS101.A4 2022 (print) | LCC RS101 (ebook) | DDC 615.1076—dc23/ eng/20211220
LC record available at https://lccn.loc.gov/2021061421
LC ebook record available at https://lccn.loc.gov/2021061422

ISBN: 978-1-032-13692-9 (hbk)
ISBN: 978-1-032-13688-2 (pbk)
ISBN: 978-1-003-23045-8 (ebk)

DOI: 10.1201/9781003230458

Typeset in Palatino
by Apex CoVantage, LLC

To

My late parents

My wife

My kids

My Father-in-Law, Mohammed Al-Seraji

My uncle and mentor Prof. Ahmed Mohammed Alhaddad

Dr. Long Chiau Ming

My mentor Prof. Nageeb Hassan

My mentor Prof. Syed Azhar Syed Sulaiman

Dr. Mohammad Bin Saqeer Alorainy

Dr. Abullah Aldahbali

My brothers: Moammer, Akram, and Ahmed

My brother Hisham Alboraihi

My brother Dr. Abdulsallam Abuelsamen

My cousin Abdulrahman Alsaar

My cousin Kamal Alsaar

My cousin Sheik Ali bin Hasan Khorsan

Contents

Preface...xi
Acknowledgments xiii
Author Bio xv

Section 1 Online Pharmacy Education.................................1

1. History and Importance 3
2. Degrees, Programs, Certificates, and Advanced
   Boards Certificates..9
3. Curriculum-Related Issues 17
4. Competencies and Learning Outcomes 23
5. Teaching the Theory 31
6. Teaching the Practice and Tutorial... 37
7. Introductory Pharmacy Practice Experiences (IPPE) and
   Advanced Pharmacy Practice Experiences (APPE).............. 45
8. Technologies and Tools 51
9. Self-Learning and Self-Directed Learning 57
10. Continuous Pharmacy Education and Professional
    Development for Pharmacy Educators.................................. 63
11. Community Services... 69
12. Access and Equitable Access 73
13. Quality and Accreditation 79
14. Advantages and Disadvantages... 85

Section 2 Online Pharmacy Practice...............................91

15. History and Importance 93
16. Online Pharmacies.. 99
17. Social Media, Social-Networking Sites, and Webinar
    and Video Conferencing Platforms 105

18. Mobile Health Technologies 111

19. Medications Safety 119

20. Patient Care 125

21. Continuing Professional Development (CPD) and
    Lifelong Learning 133

22. Pharmacists' Prescribing 143

23. Advantages, Disadvantages, and Quality Issues 153

Section 3   Online Pharmacy Research 161

24. History and Importance 163

25. Terminologies 169

26. Research Methods and Methodology 177

27. Tips for Implementation 183

28. Quality of Online Research 189

29. Facilitators and Barriers 195

Section 4   Pharmacy Education Teaching Strategies 201

30. Pharmacy Education: Learning Styles 203

31. Traditional and Active Strategies 209

32. Team-Based Learning in Pharmacy Education 217

33. Problem-Based Learning in Pharmacy Education 225

34. Case-Based Learning in Pharmacy Education 231

35. Simulation in Pharmacy Education 241

36. Project-Based Learning in Pharmacy Education 249

37. Flipped Classes in Pharmacy Education 255

38. Educational Games in Pharmacy Education 265

39. Web-Based Learning in Pharmacy Education 273

40. Lecture-Based/Interactive Lecture-Based Learning in
    Pharmacy Education 279

41. Blended Learning in Pharmacy Education 285

42. Massive Open Online Courses in Pharmacy Education 293

43. Computer-Assisted Learning and
    Computer-Based Learning in Pharmacy Education 299

Section 5 Pharmacy Education Assessment and
Evaluation Methods 305

44. Assessment Methods in Pharmacy Education: Strengths and
    Limitations 307

45. Assessment Methods in Pharmacy Education: Direct
    Assessment 317

46. Assessment Methods in Pharmacy Education: Indirect
    Assessment 325

47. Assessment Methods in Pharmacy Education: Formative
    Assessment 333

48. Objective Structured Clinical Examination (OSCE) in
    Pharmacy Education 341

Index 349

Preface

Pharmacy education has changed during the last decades in developed countries as well as in developing countries towards self-directed learning which improves the future of pharmacy professionals' and researchers' skills to be lifelong learners and to be able to use, and adapt the new technologies to improve their education, practice, research, and patient care. Furthermore, technology has played an important role in education, practice, and research and contributed effectively to pharmacy educators, students, professionals, and researchers in pharmacy education, practice, and research. This book comprises five sections containing more than 45 chapters. The five sections are as follows: Section 1 will focus on Online Pharmacy Education related issues. It includes 14 chapters. It describes the history of pharmacy education and how online education began in Chapter 1; moreover, it describes the importance of pharmacy education in general and online pharmacy education. Pharmacy/online pharmacy degrees, certificates, boards, curriculum-related issues, learning outcomes and competencies, technology and tools for effective online education and other pharmacy education, and online pharmacy education issues are covered in Chapters 2 to 14. Section 2 will focus on Online Pharmacy Practice related issues. It includes nine chapters. It describes how the history of pharmacy practice and online practice began in Chapter 15; moreover, it describes the importance of pharmacy practice and online pharmacy practice. It describes online pharmacies, online patient care, online prescribing, and other practice-related issues in Chapters 16 to 23. Section 3 will focus on Online Pharmacy Research related issues. It includes six chapters In Chapter 24, it describes the importance of pharmacy research and online pharmacy practice. It describes online research methods and other research-related issues in Chapters 25 to 29. Section 4 will focus on Pharmacy Education Teaching Strategies. It includes 14 chapters. Section 5 will focus on Pharmacy Education Assessment and Evaluation methods. It includes four chapters about evaluation and assessment methods in pharmacy education. I hope that this book can provide pharmacy educators, students, practitioners, researchers, and other readers with the necessary information and practical guidelines about online pharmacy education, practice, research, teaching strategies, and assessment methods in pharmacy education.

Yaser Mohammed Al-Worafi
December, 2021

Acknowledgments

It would have been difficult to write such a book without the help of: My wife for providing me the time and support to work on the book and spend less time with the family.

Hilary LaFoe for her valuable guide, advise and help during the writing of this book.

Acknowledgements

Author Bio

Prof. Yaser Mohammed Al-Worafi is Professor of Clinical Pharmacy at College of Pharmacy, University of Science and Technology of Fujairah, UAE (previously known as Ajman University). He graduated with a bachelor's degree in Pharmacy (BPharm) from Sana'a University, Yemen, and obtained Master and PhD degrees in Clinical Pharmacy from the Universiti Sains Malaysia (USM), Malaysia. He has more than 20 years' experience in education, practice, and research in Yemen, Saudi Arabia, United Arab Emirates, and Malaysia. He has held various academic and professional positions including Deputy Dean for Medical Sciences College; PharmD Program Director, Head of Clinical Pharmacy/Pharmacy Practice department; Head of Teaching and Learning Committee, Head of Training Committee, Head of Curriculum Committee and other committees. He has authored over 100 peer-reviewed papers in international journals and book chapters and has edited more than 10 books by Springer, Elsevier, and Taylor & Francis, USA. Prof. Yaser has supervised/co-supervised many PhD, Master, PharmD, and B-Pharm students. He is a reviewer for eight recognized international peer-reviewed journals. Prof. Yaser prepared, designed, and wrote many pharmacy programs for many universities including Master of Clinical Pharmacy/Pharmacy Practice program; PharmD program and BPharm program; internship/clerkships for Master, PharmD, and BPharm programs; more than 30 courses related to Clinical Pharmacy, Pharmacy Practice, Social Pharmacy, and Patient Care.

Yaser Mohammed Al-Worafi, PhD
Professor of Clinical Pharmacy, College of Pharmacy,
University of Science and Technology of Fujairah, Fujairah, UAE

Deputy Dean and Professor, College of Medical Sciences,
Azal University for Human Development, Sana'a, Yemen.

Section 1

Online Pharmacy Education

1

History and Importance

1.1 Pharmacy Education

The term "pharmacy education" can be used to describe: teaching pharmacy-related courses, at the undergraduate level, postgraduate level, and continuous pharmacy education level; training students at Introductory Pharmacy Practice Experiences (IPPE) and Advanced Pharmacy Practice Experiences (APPE) levels; preparing students with the essential knowledge and skills to be able to provide pharmacist care and patient care services and other services in different settings. Pharmacy education has changed during the last decades in developed countries as well as in developing countries towards self-directed learning which improves the future of pharmacy professionals' and researchers' skills to be lifelong learners and to be able to use and adapt the new technologies to improve their education, practice, research, and patient care. Furthermore, technology has played an important role in education, which has led to replacing or decreasing the need for traditional laboratories such as adapting simulation for pharmacology, anatomy, and other subjects, which has led to improving learning outcomes and, moreover, saving the pharmacy schools budgets. Moreover, promoting interprofessional and intraprofessional education has been implemented successfully in many schools to prepare students/future health care professionals with the essential knowledge, skills, and attitudes for effective collaborative interprofessional/intraprofessional practice. Pharmacy practices in developed and many developing countries have witnessed many changes during the last decades towards improving pharmacist care/patient care services and pharmacy schools/departments have updated/reformed the curriculum in order to fill the gap between practice and education, to improve the competencies of graduates and students, and to contribute effectively to patient care and practice. Pharmacy education and practice plays a very important role worldwide and provides very important services to students, the public, patients, health care providers, and policy makers.

DOI: 10.1201/9781003230458-2

1.2 History of Pharmacy Education

It is reported that pharmacy practice history goes back to ancient times, when old men used natural resources such as herbs for their health. However, the history of pharmacy education or what was previously called "pharmaceutical education" goes back to the 1700s when the first college of pharmacy was formed with the founding of the "Collège de pharmacie" in 1771, France (Warolin, 2003).

It was the 1800s when the first college of pharmacy was founded in the United States in 1821, and is now known as the Philadelphia College of Pharmacy and Science (Fink, 2012; AIH, 2018). It has been reported that pharmacy was considered an art and not a science in the 1700s (AIH, 2018). Historical records reported the following examples of development in pharmacy education around the world (Warolin, 2003; Fink, 2012; AIH, 2018; Cox, 2019; Holloway, 1995; Elsayed et al., 2016; Ibrahim et al., 2016; Basak and Sathyanarayana, 2010; Khachan et al., 2010):

Pharmacy was considered an art and not a science in the 1700s.

In 1777 was the founding of the Collège de pharmacie (France).

Pharmacy education was based on the apprenticeship model. Most physicians provided a shop practice with an employee apothecary and/or apprentice.

Medical education in USA was similar to pharmacy education with only four medical schools prior to 1800:

1765 College of Philadelphia (University of Pennsylvania); 1767 King's College (Columbia University); 1782 (Harvard University); and 1797 (Dartmouth University).

1821, Philadelphia College of Pharmacy was established in the USA.

1823, Massachusetts College of Pharmacy was established in the USA.

1829, College of Pharmacy of the City of New York was established in the USA.

1824, the first modern pharmacy program offered in Egypt dates back to 1824, when the ruler of Egypt, Mohamed Ali Pasha, founded the Medicine and Pharmacy School as part of a hospital established in Abu Zaabal, in the province of Cairo. It was then transferred to the Citadel area in 1829 and then to Kasr El-Aini street, its current situation, in 1837.

1936, The first pharmacy college, the Royal College of Pharmacy and Chemistry, was established in 1936 in Baghdad and was affiliated with the University of Baghdad, the largest university in Iraq.

1840, Maryland College of Pharmacy was established in the USA.

1842, The pharmacy division was founded by the Royal Pharmaceutical Society of Great Britain in 1842 as the College of the Pharmaceutical Society.

1850, Cincinnati College of Pharmacy was established in the USA.

1881, Pharmacy education in Lebanon goes back to 1881 when the French Jesuit missionaries established Université Saint Joseph (USJ), then USJ founded the first formal pharmacy education program in Lebanon in 1912.

1932, Prof. Mahadev Lal Schroff (called the Father of Pharmacy Education in India) started a Pharmacy college/department at the Banaras Hindu University; formal pharmacy education leading to a degree began with the introduction of a 3-year Bachelor of Pharmacy (BPharm) at Banaras Hindu University in 1937. The first independent Pharmacy College in was started in Goa–India, in 1842, by the Portuguese.

1.3 Distance Education

The term "distance education" can be used to describe teaching students outside the universities and colleges when they can't attend the lectures at the university and college campuses due to many factors such as geographic reasons, financial reasons, and others. Therefore, the students will learn/study at their home, the materials will be received by mail, or other methods of delivery, and by the help of technologies/new technologies through the last decades. Distance education history goes back to 1728, when Caleb Phillips, teacher of the new method of "Short Hand," USA sought students who wanted to learn through weekly mailed lessons. This was followed by the first distance education course in the modern sense, provided by Sir Isaac Pitman in the 1840s, "who taught a system of shorthand by mailing texts transcribed into shorthand on postcards and receiving transcriptions from his students in return for correction. The element of student feedback was a crucial innovation in Pitman's system. This scheme was made possible by the introduction of uniform postage rates across England in 1840" (Wikipedia, Distance Education). Distance learning degrees were established in 1858 at the University of London as part of its external program offering. Electronic distance learning was started in the 1920s, 1930s, and forward by using films, radio, and television which allowed universities and schools to implement broadcast educational programs for public schools (Wikipedia, Distance Education).

1.4 Phases of Distance Education Development

Phase I. Distribution of printed teaching material through the mail through the 18th and 19th centuries.

Phase II. Broadcast of educational courses and materials through films, radio, nd television from the 1920s and forward.

Phase III. Computer assisted learning from the 1980s and forward.

Phase IV. Internet assisted technologies from the 1980s and forward.

1.5 Distance and Online Pharmacy Education

It was very difficult to adapt and implement the distance education model in pharmacy education as well as other medical and health sciences programs as the nature of the study at pharmacy, medical, and health schools requires many hours of laboratories and training at schools, primary health care facilities, secondary health care facilities, and tertiary health care facilities.

1.6 History of Distance and Online Pharmacy Education

Ward et al. (2003) reported that Southeastern University of the Health Sciences had developed a North Miami site to service its nontraditional, post baccalaureate, Doctor of Pharmacy (PharmD) program in 1991. Due to the students' inability to disrupt their careers and relocate for the purpose of proximity, program enrollment numbers were low. As a result of the request by the American Council on Pharmaceutical Education to maintain nontraditional PharmD programs, the university's existing infrastructure, and NSU's (Nova University and Southeastern University) mission to utilize technologically innovative delivery systems to provide convenient, high-quality education, a distance learning format was needed. Compressed videoconferencing was chosen over satellite broadcast as the delivery system for the PharmD program for two essential reasons: its ability to conduct high quality, live, two-way interaction at a much lower cost and NSU's previous success with this technology in the education department (Ward et al., 2003). Since the 2000s, as a result of internet technologies, many pharmacy schools worldwide launched distance online programs for postgraduate studies and continuous medical/pharmacy education.

1.7 Importance of Online Pharmacy Education

The importance of distance and online education has increased during the last decades, especially after 2019. Online education and learning had/has great impact on the continuity of the teaching and learning cycle in pharmacy worldwide; saved the pharmacy schools from closing for at least 1 year because of lockdown after the COVID-19 pandemic; nowadays, technologies such as online education platforms, social media, and mobile technologies can help to achieve the pharmacy programs' learning objectives/outcomes and provide efficient and convenient ways to achieve learning goals for pharmacy education. Online pharmacy education postgraduate programs are very important for working pharmacists as they can save money and time and achieve their degrees while they are working because it will be flexible, easy, and convenient for them. Online pharmacy education courses can help registered pharmacists to obtain the required continuous medical hours (CMEs), without being absent from their work as they can take it during their weekends and off hours with less cost. Online pharmacy courses and programs can help pharmacy schools worldwide to increase their income with less cost.

Conclusion

This chapter has discussed the history of distance education in general as well as in pharmacy education, phases of distance education and learning education development through history, and the importance of distance learning education in pharmacy education. Online pharmacy education nowadays plays an important role in the pharmacy education teaching and learning cycle and helps pharmacy educators, pharmacy students, and schools to achieve the pharmacy education programs' learning goals, objectives, and outcomes.

References

AIHP., 2018. *Guidelines on teaching history in pharmacy education*. Available at: https://aihp.org/wp-content/uploads/2018/12/AACP-D.pdf

Basak, S.C. and Sathyanarayana, D., 2010. Pharmacy education in India. *American Journal of Pharmaceutical Education*, 74(4).

Cox, N., 2019. Four pharmacy education entrepreneurs in Victorian Britain: Robert Clay (1792–1876), John Abraham (1813–1881), John Muter (1841–1911) and George Wills (1842–1932). *Pharmaceutical Historian*, 49(3), pp. 74–82.

Elsayed, T.M., Elsisi, G.H. and Elmahdawy, M., 2016. Pharmacy practice in Egypt. In *Pharmacy practice in developing countries* (pp. 291–317). Academic Press.

Fink, J. L., III. (2012). Pharmacy: A brief history of the profession. In *The student doctor network*. Available at: www.studentdoctor.net/2012/01/pharmacy-a-brief-history-of-the-profession/.

Holloway, S.W.F., 1995. *Producing experts, constructing expertise: The school of pharmacy of the pharmaceutical society of Great Britain, 1842–1896* (pp. 116–140). Brill Rodopi.

Ibrahim, I.R. and Wayyes, A.R., 2016. Pharmacy practice in Iraq. In *Pharmacy practice in developing countries* (pp. 199–210). Academic Press.

Khachan, V., Saab, Y.B. and Sadik, F., 2010. Pharmacy education in Lebanon. *Currents in Pharmacy Teaching and Learning*, 2(3), pp. 186–191.

Ward, C.T., Rey, J.A., Mobley, W.C. and Evans, C.D., 2003. Establishing a distance learning site for a traditional doctor of pharmacy program. *American Journal of Pharmaceutical Education*, 67(1/4), p. 153.

Warolin, C., 2003. The foundation of the school of pharmacy in Paris. *Revue D'histoire de la Pharmacie*, 51(339), pp. 453–474.

2

Degrees, Programs, Certificates, and Advanced Boards Certificates

2.1 Pharmacy Education Degrees

Pharmacy education schools/departments offer the following degrees:

Diploma in Pharmacy

Study 2 to 3 years; it may be different from one country to another. Then graduates can work as pharmacy technicians or assistants. Licensing could be different from one country to another.

Bachelor of Pharmacy or Pharmaceutical Sciences

Study 3 to 6 years with or without foundation; it may be different from one country to another. Then graduates can work in different places such as pharmacists in community pharmacies, hospitals, pharmaceutical industries, companies, research centers, drug authorities, and others. Licensing could be different from one country to another.

Doctor of Pharmacy (PharmD)

Study duration is affected by the nature of PharmD program-related factors such as: undergraduate or postgraduate program; with or without foundation; traditional or non-traditional program and others. It may be different from one country to another. Then graduates can work in different places such as pharmacists in the community pharmacies, hospitals, pharmaceutical industries, universities, companies, research centers, drug authorities, and others. Licensing could be different from one country to another.

Postgraduate Diploma

Study 1 to 2 years. It can help students to prepare for their master studies.

Master's in Pharmacy

Study duration is affected by the nature of the program and study mode such as: research/project based only, course based, mixed mode (courses and research project), full or part time, and others.

Doctorate or Doctor of Philosophy (PhD)

Study duration is affected by the nature of program and study mode such as: research/project based only, course based, mixed mode (courses and research project), full or part time, and others.

2.2 Online Pharmacy Education Degrees

It was very difficult to adapt and implement the distance education model in pharmacy education undergraduate programs as the nature of the study at pharmacy schools' programs required many laboratories, as well as training at schools, primary health care facilities, secondary health facilities, and tertiary health care facilities. However, after the new technologies were successfully implemented in the early 1990s by the Southeastern University of the Health Sciences, which developed a North Miami site to service its non-traditional, post baccalaureate, Doctor of Pharmacy (PharmD) program in 1991 and forward worldwide. Online pharmacy education has changed towards adapted and implemented programs, and therefore offers online degrees for pharmacy students, especially for postgraduate studies. Since early 2020, the majority of pharmacy schools/departments have offered all degrees at both undergraduate and postgraduate levels completely online or as a hybrid model (theory courses delivered online, training in the hospitals or pharmacies).

2.3 Pharmacy Education Programs

Pharmacy education schools/departments offer the following degrees:

Undergraduate Programs

Bachelor of Pharmacy (B-Pharm), Bachelor of Pharmaceutical Sciences (BSc), Bachelor of Clinical Pharmacy, Doctor of Pharmacy (PharmD).

Postgraduate Programs

Doctor of Pharmacy (PharmD):

Study is traditional on the campus or in a non-traditional program for working pharmacists which allows them to study the theory courses completely online.

Postgraduate Diploma

Study 1 to 2 years. It can help students to prepare for their master's studies.

At Master Level

Many programs with or without subspecialty such as:

Clinical Pharmacy; Pharmacy Practice; Master of Science; Pharmaceutical Care; Pharmacology (Applied Clinical); Pharmacognosy; Medicinal/Pharmaceutical Chemistry; Pharmaceutics; Pharmacovigilance; Pharmacoepidemiology; Social Pharmacy; Administrative Pharmacy; and others. Master's programs can be different from one country to another. Pharmacy students can also study for their master's at other schools such as a school of medicine.

At Doctorate/PhD Level

Many programs with or without subspecialty such as:

Clinical Pharmacy; Pharmacy Practice; Master of Science; Pharmaceutical Care; Pharmacology (Applied Clinical); Pharmacognosy; Medicinal/Pharmaceutical Chemistry; Pharmaceutics; Pharmacovigilance; Pharmacoepidemiology; Social Pharmacy; Administrative Pharmacy, and others. Master's programs can be different from one country to another. Pharmacy students can also study for their master's at other schools such as a school of medicine.

2.4 Online Pharmacy Education Programs

Nowadays after the successful implementation of the new technologies in the early 1990s in pharmacy education, pharmacy schools/departments can offer many postgraduate programs completely online, especially postgraduate diploma and master's programs.

2.4.1 Pharmacy Education Certificate Courses and Programs

Certificate courses are very important for pharmacists, educators, and students to improve their practical knowledge and skills. Nowadays many local and international universities, organizations, and publishers such as Elsevier provide many certified courses with cost as well as free courses. The duration of certified courses can be days, weeks, or months depending on the type of certified course and program.

2.4.2 Online Pharmacy Education Certificate Courses and Programs

Pharmacists can register for many certificate courses or programs online, saving time and cost. Examples of Certificate Courses and Programs are:

Anticoagulation Certificate

Organized by the American Society of Health-System Pharmacists (ASHP): 32 continuous education (CE) hours designed for pharmacists to expand their knowledge and skills in anticoagulation therapy management across special populations and within acute care, ambulatory, and perioperative settings (ASHP, 2021).

Pharmacogenomics Certificate

Organized by the American Society of Health-System Pharmacists (ASHP): 20 continuous education (CE) hours designed for participants to increase the knowledge and skills necessary to use pharmacogenomics to improve medication use in a variety of patient care settings (ASHP, 2021).

Pain Management Certificate

Organized by the American Society of Health-System Pharmacists (ASHP): 21.5 continuous education (CE) hours designed for pharmacists to develop the knowledge and skills necessary to provide optimal pain management in patients suffering from chronic pain (ASHP, 2021).

Diabetes Management Certificate

Organized by the American Society of Health-System Pharmacists (ASHP): 33 continuous education (CE) hours designed to increase the knowledge and skills associated with the diagnosis, management, and pharmacological treatment of diabetes to optimize patient outcomes in ambulatory and inpatient care settings (ASHP, 2021).

Medication Safety Certificate

Organized by the American Society of Health-System Pharmacists (ASHP): 40 continuous education (CE) hours designed to enhance the skills and knowledge for pharmacy professionals, physicians, and nurses who lead medication safety improvements (ASHP, 2021).

2.4.3 Advanced Certificates and Board Certificates for Pharmacists

There are many advanced certificates and board certificates for pharmacists organized by many national and international organizations. Certain courses are equal/accredited as professional master's degree in many countries such as "Board Certified Pharmacotherapy Specialist (BCPS)."

The history of board certificates goes back to the early 1970s by the American Pharmacists Association (APhA).

Advanced certificates and board certificates are contributing effectively to the improvement of pharmacist care/patient care services and of pharmacists' knowledge and skills towards advanced services. Examples of Advanced Certificates and Board Certificates are:

The Pharmacist Independent Prescribing Practice Certificate

The Pharmacist Independent Prescribing Practice Certificate is provided by many universities in the United Kingdom (UK). This program/certificate aims to improve the registered pharmacist's knowledge and skills towards prescribing and prepare them to be good pharmacist prescribers.

Board Pharmacy Specialties (BPS)

The BPS Board Certified Nuclear Pharmacist specializes in the procurement, preparation, compounding, dispensing, and distribution of radiopharmaceuticals, as well as the regulatory aspects governing these processes. In addition, the nuclear pharmacist serves as the medication expert within the health care team regarding clinical aspects of radiopharmaceuticals and non-radioactive drugs used in patient care (BPS, 2021).

Board Certified Medication Therapy Management Specialist (BCMTMS)

Board Certified Medication Therapy Management Specialists (BCMTMS) are organized by the National Board of Medication Therapy Management (NBMTM), designed to improve the pharmacist's knowledge and skills towards Medication Therapy Management and prepare them to provide effective Medication Therapy Management for patients.

Oncology Pharmacy Board Certificate

Oncology Pharmacy Board Certificates provide evidence-based, patient-centered medication therapy management and direct patient care for individuals with cancer, including treatment assessment and monitoring for potential adverse drug reactions and interactions (BPS, 2021).

Solid Organ Transplantation Pharmacy Certificate

The Solid Organ Transplantation Pharmacy Certificate provides evidence-based, patient-centered medication therapy management and care for patients throughout all phases of solid organ transplantation at all ages and in various health care settings (BPS, 2021).

Ambulatory Care Pharmacy Certificate

The Ambulatory Care Pharmacy Certificate addresses the provision of integrated, accessible health care services for ambulatory patients in a wide variety of settings, including community pharmacies, clinics, and physicians' offices (BPS, 2021).

Compounded Sterile Preparations Pharmacy Certificate

The Compounded Sterile Preparations Pharmacy Certificate ensures that sterile preparations meet the clinical needs of patients, satisfying quality, safety, and environmental control requirements in all phases of preparation, storage, transportation, and administration in compliance with established standards, regulations, and professional best practices (BPS, 2021).

2.5 Conclusion

This chapter has discussed the online pharmacy degrees, programs, certificates, advanced certificates, and board certificates and their impact on pharmacist care and patient care services. Online pharmacy education has contributed effectively to the delivery of pharmacy programs and courses.

References

ASHP., 2021. *Professional certificates.* Available at; www.ashp.org/professional-development/professional-certificate-programs?loginreturnUrl=SSO CheckOnly

BPS., 2021. www.bpsweb.org/bps-specialties/nuclear-pharmacy/#1517761118361-6c02bae3-f5a01517848269340151785128139 5

3

Curriculum-Related Issues

3.1 Pharmacy Education Curriculum

The pharmacy education program's curriculum has changed many times during the last decades because of the development of the pharmacy profession. Pharmacy education or what was previously called "pharmaceutical education" goes back to the 1700s when the first college of pharmacy was founded as the "Collège de pharmacie" in 1771, France (Warolin, 2003), followed by many schools and programs around the world (Please see Chapter 1 for more details). Pharmacy education curriculum was focused on herbals, manufacturing, and simply in the products rather than patient outcomes until the revolutionary change in the pharmacy practice a few decades ago, when the pharmacy profession witnessed great practice changes, and moved away from its original focus on medicine supply and dispensing towards a focus on patient care, especially after the introduction of clinical pharmacy concepts in the late 1960s, followed by the philosophy of pharmaceutical care in the early 1990s. However, pharmacists nowadays play/should play an important role in patient care. They contribute effectively to patients' health, diseases/conditions management, as well as prevention; treating outcomes; treating cost; patients' quality of life; satisfaction towards the health care system and care. Therefore, pharmacy education should prepare future pharmacists with the essential knowledge and skills to be good pharmacists able to provide effective pharmacist care/patient care services.

3.2 Phases of Pharmacy Education Curriculum Development

Literature reported the following phases in the development of pharmacy education curriculum throughout history (Parascandola, 1995; Parascandola, 2000; Zunic et al., 2017; Cahlíková et al., 2020; Cubukcu, 1998; Dhami, 2013; Manasse and Rucker, 1984; Turner, 1986; Miller, 1981):

DOI: 10.1201/9781003230458-4

Phase I. Beginning of Pharmacy/Pharmaceutical Education

In the 1700s, pharmacy education was focused only on chemistry and herbs.

Phase II. Introducing Bacteriology, Pharmacology, and Physiological Chemistry

The second half of the 19th century was a period of revolutionary change in the biomedical sciences. The development of the germ theory of disease and the establishment of fields such as bacteriology, pharmacology, and physiological chemistry laid the groundwork for significant advances in medicine and pharmacy. The extraordinary growth of the biomedical sciences also made its impact felt on medical and pharmaceutical education.

Phase III. Introducing Laboratory Training

The pharmacy program established at the University of Michigan in 1868, under the leadership of physician-chemist Albert Prescott, made the study of pharmacy essentially a full-time occupation for the 2 years of the curriculum. It also introduced extensive laboratory training in the basic sciences. Other schools of pharmacy eventually adopted this model, beginning with the University of Wisconsin, which established its pharmacy program in 1883.

Phase IV. Introducing Pharmacognosy

Since the 1800s chemistry and pharmacognosy have dominated the pharmaceutical sciences in colleges of pharmacy. In the past, traditional medicinal knowledge prevalent in the form of holy books, incantations, folklores, Materia Medica, and other historical literature defined the preliminary guidelines for the authorization of plant-derived natural medicines. The conventional medical practices adopted for identification and authentication of natural remedies eventually framed the botanico-chemical approach to pharmacognosy during the early 19th century. However, the last 200 years witnessed a substantial metamorphosis in the principles and practices of pharmacognosy and it has become an essential domain of modern pharmaceutical science.

Phase V. Introducing Pharmaceutical Sciences

Pharmaceutical sciences courses were introduced and implemented after the evolution of pharmaceutical industries, where research laboratories became common in the major drug firms in the late 19th century.

Phase VI. Introducing Clinical and Pharmaceutical Care Pharmacy Sciences

Clinical pharmacy and pharmaceutical care courses have been implemented since the 1960s.

Phase VII. Introducing Social, Administrative, and Economic Sciences

Social, administrative, and economic related courses have been implemented since the 1970s in Europe as well as around the world.

Phase VIII. Introducing Online Courses

After the successful implementation of new technologies in the early 1990s, the Southeastern University of the Health Sciences, which developed a North Miami site to service its non-traditional, post baccalaureate, Doctor of Pharmacy (PharmD) program in 1991 and forwards worldwide, online pharmacy education has changed to adapt and implement many programs, and therefore offer online degrees for pharmacy students, especially for postgraduate studies. Since early 2020, the majority of pharmacy schools/ departments have offered all degrees at both undergraduate and postgraduate levels completely online or as a hybrid model (theory courses delivered online with training in hospitals or pharmacies).

3.3 Online Curriculum Reform Related Issues

There are many issues that pharmacy schools/departments should plan for carefully to deliver the programs and courses online effectively and achieve the program/course goals, objectives, and learning outcomes as follows:

Identify and revise the goals, needs, and competencies.

Identify and revise the learning outcomes.

Identify and revise, organize, and design the courses.

Identify and revise the course credit hours.

Identify and revise the theory, practical, tutorial, and training for all courses and clerkships.

Design and revise the course objectives, learning outcomes, teaching strategies, assessments, and evaluation methods.

Link the course learning outcomes with the program learning outcomes.

Identify the appropriate delivery methods.

Prepare the facilities and resources for educators and students such as technology, media, manuals, and others.

Conduct workshops and training for educators and students.

Obtain regular feedback from educators, students, and staff.

Identify the actual and potential problems, design and plan to solve them and avoid them in the future.

Review the curriculum regularly and send it for peer/expert review outside your university.

3.4 Achievements

Nowadays, the majority of pharmacy schools/departments in developed countries as well as in many developing countries have implemented and delivered the pharmacy curriculums for undergraduate and postgraduate programs online successfully. Pharmacy schools/departments nowadays are able to design and implement pharmacy program curriculums better than at any time in the history.

3.5 Challenges and Recommendations

There are many challenges affecting the successful implementation of pharmacy program curriculums such as the following:

Practical/Laboratory Related Courses

Teaching practical/laboratory as part of many courses through the pharmacy education curriculum at both undergraduate and postgraduate levels is very important to achieve the course and program objectives and learning outcomes. However, there are many effective strategies to overcome this challenge such as: adapting simulation as a teaching strategy could overcome and solve this challenge for many courses; teaching the practical/laboratory at pharmacy schools/departments labs in one week or more instead of once weekly through the semester and apply the local guidelines regarding the lab size and so on.

Introductory Pharmacy Practice Experiences (IPPE) and Advanced Pharmacy Practice Experiences (APPE)

IPPE and APPE in the pharmacy education curriculum are very important as an essential part of any undergraduate curriculum worldwide to prepare students with the essential knowledge and skills to be able to provide pharmacist care and patient care services and other services in different settings. However, there are many effective strategies to overcome this challenge such as: adapt simulation training to overcome and solve this challenge for many IPPE and APPE clerkships; however, schools/departments can also adapt the hybrid model for the training and combine simulation with onsite training.

Online Teaching and Assessment

Lack of facilities to move towards online teaching and assessment is a very important challenge facing many pharmacy schools/departments, especially in developing countries. However, there are many effective strategies to overcome and solve this challenge such as recording the classes and mailing/emailing them to the students. Change the assessment methods according to the available facilities/resources in the schools/departments.

Extracurricular Activities

Extracurricular activities are associated with many benefits to pharmacy students such as social benefits, cognitive benefits, well-being benefits, and other benefits. By the nature of online education, extracurricular activities could be affected and considered as a challenge to the students as well as pharmacy schools/departments. However, adapting social media and new technologies can help to overcome and solve this challenge.

Technologies

Technologies facilities are very important in pharmacy education. Lack of essential technologies is one of the major challenges to students, educators, and schools/departments around majority/many middle/low income developing countries. However, adapting the old distance education tools can help to overcome this challenge.

Quality and Accreditation

Lack of quality standards and guidelines in many countries about online education is considered a very important challenge. Lack of monitoring in online pharmacy education can affect the competencies of the future pharmacists and therefore, affect patient care and health care. However, developing, adapting, and implementing quality standards, guidelines, and effective monitoring systems can overcome and solve this challenge.

3.6 Conclusion

This chapter has discussed the development of pharmacy curriculums throughout history. This chapter includes background about pharmacy education curriculums, development, and phases. It also describes the achievements in the online curriculum, as well as challenges and suggests recommendations to overcome and solve the identified challenges. Pharmacy schools/departments nowadays are able to design and implement pharmacy program curriculums better than at any time in the history.

References

Cahlíková, L., Šafratová, M., Hošťálková, A., Chlebek, J., Hulcová, D., Breiterová, K. and Opletal, L., 2020. Pharmacognosy and its role in the system of profile disciplines in pharmacy. *Natural Product Communications*, 15(9), p. 1934578X20945450.

Cubukcu, B., 1998. A brief history of the development of Turkish pharmacognosy. Yeni tip tarihi arastirmalari. *The New History of Medicine Studies*, 4, pp. 225–232.

Dhami, N., 2013. Trends in pharmacognosy: A modern science of natural medicines. *Journal of Herbal Medicine*, 3(4), pp. 123–131.

Manasse, H.R. and Rucker, T.D., 1984. Pharmacy administration and its relationship to education, research and practice. *Journal of Social and Administrative Pharmacy*, 2, pp. 127–35.

Miller, R.R., 1981. History of clinical pharmacy and clinical pharmacology. *The Journal of Clinical Pharmacology*, 21(4), pp. 195–197.

Parascandola, J., 1995. The emergence of pharmaceutical science. *Pharmacy in History*, 37(2), pp. 68–75.

Parascandola, J., 2000. The pharmaceutical sciences in America, 1852–1902. *Journal of the American Pharmaceutical Association*, 40(6), pp. 733–735.

Turner, P., 1986. The nuffield report: A signpost for pharmacy. *British Medical Journal* (Clinical research ed.), 292(6527), p. 1031.

Zunic, L., Skrbo, A. and Dobraca, A., 2017. Historical contribution of pharmaceutics to botany and pharmacognosy development. *Materia Socio-Medica*, 29(4), p. 291.

4

Competencies and Learning Outcomes

4.1 Pharmacy Education Competencies

Competencies (the ability to do things well) in pharmacy education go back to the 1700s when the first college of pharmacy was founded as the "Collège de pharmacie" in 1771, France (Warolin, 2003) and developed through the centuries according to the development of pharmacy profession and practice. However, pharmacists nowadays play/should play an important role in patient care, and they contribute effectively to health, diseases/conditions management as well as prevention; treating outcomes; treating cost; patients' quality of life; satisfaction towards the health care system and care. Pharmacists nowadays provide online pharmacist care services in many countries around the world. Therefore, pharmacy schools/departments should review and update the pharmacy education programs' competencies and learning outcomes regularly to be able to graduate students with the essential knowledge and skills towards providing effective online services. Many schools since the 1970s have implemented competency-based pharmacy education (Knapp and Supernaw, 1977).

4.2 Competencies-Based Education Principles

Competencies-based education has the following principles (Frank et al., 2010):

Focusing on Outcomes

To be ensure that all graduates are competent in all essential domains. Prepare the graduates to practice, to be good professionals able to provide effective patient care services.

DOI: 10.1201/9781003230458-5

Emphasizing Abilities

Medical curricula must emphasize the abilities to be acquired. An emphasis on the abilities of learners should be derived from the needs of those served by graduates (i.e., societal needs).

De-Emphasizing Time-Based Training

Medical education can shift from a focus on the time a learner spends on an educational unit to a focus on the learning attained. Greater emphasis should be placed on the developmental progression of abilities and on measures of performance.

Promoting Greater Learner-Centeredness

Medical education can promote greater learner engagement in training. A curriculum of competencies provides clear goals for learners.

4.3 Pharmacy Education/Online Pharmacy Education Competencies and Learning Outcomes

Local and international accreditations agencies are responsible for the development, updating, and approving of the competencies and learning outcomes for pharmacy education programs to prepare the future pharmacists to work locally and internationally, taking into consideration what the country needs as well as how to work outside the country. Nowadays, pharmacy education competencies and learning outcomes are well designed, better than at any time in history. Examples of competencies and learning outcomes are developed based on the literature plus personal experience and opinion (ACPE, 2015; GPS, 2012; NAPRA/ANORP, 2014; Rouse and Meštrovic, 2014; Sacre et al., 2020).

The future (graduates) pharmacists should be able to:

Knowledge (Learner, Educator, and Health and Wellness Promoter and Counselor)

- Describe the basics/principles/fundamentals of biomedical sciences, pharmaceutical sciences, clinical, social/administrative/behavioral sciences, and clinical sciences.
- Describe the diseases, conditions, medicines information, management plan (goals of therapy and desired outcomes, non-pharmacological interventions, pharmacological interventions,

monitoring parameters), and health-related issues to patients, as well as the public and health care professionals.

- Demonstrate ability to learn online when needed.

Cognitive (Thinker, Analyzer, Problem Solver, and Decision Maker)

- Apply the biomedical sciences, pharmaceutical sciences, clinical, social/administrative/behavioral sciences, and clinical sciences knowledge in pharmacist care/patient care services and activities.
- Apply critical thinking to identify, solve, and minimize drug-related problems (DRPs).
- Retrieve and evaluate drug information from pharmaceutical and biomedical science resources and reports for application in specific patient care situations to enhance clinical decision making.
- Apply basic scientific pharmaceutical and clinical knowledge in calculations and solving clinical cases.

Communication, Education, and Collaboration (Communicator, Educator, and Collaborator)

- Demonstrate effective communication with patients, the public, health care professionals, students, organizations, and societies face to face or online if needed.
- Demonstrate ability to work effectively within teams (interprofessional and intraprofessional).
- Demonstrate ability to educate, counsel patients, the public, health care professionals and students about their diseases, conditions, medicines information, management plan (goals of therapy and desired outcomes, non-pharmacological interventions, pharmacological interventions, monitoring parameters), and health-related issues face to face or online if needed.

Life-Long Learning and Personal/Professional Development (Learner and Innovator)

- Demonstrate ability to take responsibility towards their learning.
- Demonstrate ability to engage in innovative activities by using creative thinking to envision better ways of accomplishing professional goals.
- Demonstrate ability to use online sources/resources for their learning.

Leadership and Management (Leader)

- Demonstrate ability to lead teams effectively (interprofessional and intraprofessional).
- Demonstrate leadership abilities in professional endeavors.
- Demonstrate effective leadership and management skills as part of the multi-disciplinary team.
- Take appropriate actions to respond to complaints, incidents, or errors in a timely manner and prevent them from happening again.
- Demonstrate resilience and flexibility, and apply effective strategies to manage multiple priorities, uncertainty, complexity, and change.
- Develop, lead, and apply effective strategies to improve the quality of care and safe use of medicines.

Pharmaceutical Marketing, Pharmacist Care Marketing (Promoter)

- Describe the concept of pharmaceutical marketing.
- Describe the various components of promotion of pharmaceutical products.
- Demonstrate ability to market pharmaceutical products and cosmetics.
- Demonstrate ability to market pharmacist care services and activities.
- Demonstrate ability to use different technologies for the purpose of marketing.

Pharmaceutical Industry (Manufacturers)

- Demonstrate ability to perform pharmaceutical compounding.
- Demonstrate ability to work effectively in the pharmaceutical industries.
- Demonstrate ability to use different and new technologies in the pharmaceutical industries.

Pharmacist and Patient Care (Care Provider and Innovator)

- Demonstrate ability to provide effective pharmacist care services to patients, the public, students, and health care professionals face to face or online if needed.
- Apply pharmacist care to achieve and improve the clinical, economical, and humanistic outcomes for treating patients and public health diseases and conditions.

- Demonstrate ability to recognize social determinants of health to diminish disparities and inequities in access to quality care.
- Demonstrate ability to represent the patient's best interests.
- Demonstrate effective communication with patients, the public, and health care professionals during pharmacist/patient care services.
- Demonstrate ability to be creative and an innovator towards patient care.

Medication Safety (Care Provider and Health Protector)

- Demonstrate ability to dispense medications, herbal medication, and other nutraceuticals appropriately, safely, and effectively to patients and the public.
- Demonstrate ability to perform pharmaceutical compounding and patient-specific calculations, including pharmacokinetic and other therapeutic calculations to individualize the dosage regimens for patients when needed.
- Demonstrate ability to identify the potential drug-related problems (DRPs) such as potential adverse drug reactions (ADRs), to minimize them as well as actual DRPs such as ADRs and manage them.
- Demonstrate ability to work effectively with other health care professionals to minimize, prevent, and manage actual medication errors, and to prevent/minimize potential medication errors.
- Demonstrate ability to improve knowledge and skills towards medication safety aspects.
- Demonstrate ability to collaborate effectively in developing, implementing, and evaluating policies, procedures, and activities that promote quality and safety.

Prescribing (Prescriber)

- Demonstrate ability to provide safe and effective prescribing based on evidence-based medicine.
- Demonstrate ability to develop a systematic, evidence-based, and reflective approach to prescribing practice.
- Demonstrate ability to provide safe and effective prescribing and consultations online.

Ethical, Legal, and Professional Responsibilities

- Demonstrate integrity, honesty, knowledge of ethical principles and the standards of professional conduct, and the ability to apply ethical principles in pharmacist care/patient care, research, education, or community service.

* Demonstrate ability to work within country legal requirements.

Health Promotion and Community Services

* Demonstrate ability to engage in health promotion and community services activities face to face or online.

Research (Researcher)

* Demonstrate ability to design research proposals.
* Demonstrate ability to design, implement, and conduct research as well as online research.
* Demonstrate ability to take part in research activities, audit, service evaluation, and quality improvement, and demonstrate how these are used to improve care and services.

Technology (IT User)

* Demonstrate ability to use different software/programs, mobile applications, social media, online platforms, media technologies, and other technologies for the purpose of learning, research, and practice.

4.4 Achievements

Nowadays, the majority of pharmacy schools/departments in developed countries as well as in many developing countries have adapted the competency-based/learning outcome-based education for pharmacy programs successfully.

4.5 Challenges and Recommendations

During the curriculum reform in many developing countries, pharmacy schools/departments adapted the international competencies/learning outcomes and added many missing important competencies related to the need of their countries. However, meeting with all stakeholders, pharmacy educators, practice leaders, and experts can help to overcome and solve this challenge and lead to developing, revising, and updating the competencies and leaning outcomes for the benefit of students/future students as well as countries. Elaborating on the competencies and learning outcomes related to online/distance education is very important and needed. Local and international accreditation play an important role in revising and updating the competencies and learning outcomes.

4.6 Conclusion

This chapter has discussed the development of competencies in pharmacy education throughout history. This chapter includes the history of pharmacy education competencies. It also describes the principles of competency-based education. It suggests competencies and learning outcomes for online pharmacy education, describes the achievements and challenges, and suggests recommendations to overcome and solve the identified challenges.

References

Accreditation Council for Pharmacy Education., 2015. *Accreditation standards and key elements for the professional program in pharmacy leading to the doctor of pharmacy degree* (Standards 2016). Available at: www.acpe-accredit.org/pdf/Standards2016FINAL.pdf

General Pharmaceutical Council (GPC), 2021. *Standards for pharmacy education.* Available at: www.pharmacyregulation.org/sites/default/files/document/standards-for-the-initial-education-and-training-of-pharmacists-january-2021_0.pdf

Frank, J.R., Snell, L.S., Cate, O.T., Holmboe, E.S., Carraccio, C., Swing, S.R., Harris, P., Glasgow, N.J., Campbell, C., Dath, D. and Harden, R.M., 2010. Competency-based medical education: Theory to practice. *Medical Teacher*, 32(8), pp. 638–645.

Knapp, K.K. and Supernaw, R.B., 1977. A systematic approach to the development of a competency-based doctor of pharmacy program. *American Journal of Pharmaceutical Education*, 41(3), pp. 290–295.

NAPRA (National Association of Pharmacy Regulatory Authorities)/ANORP (Association nationale des organismes de réglementation de la pharmacie)., 2014. *Professional competencies for canadian at entry to practice pharmacists.* Available at: https://napra.ca/sites/default/files/2017-08/Comp_for_Cdn_PHARMACISTS_at_EntrytoPractice_March2014_b.pdf

Rouse, M. and Meštrovic, A., 2014. *Quality assurance of pharmacy education: The FIP global framework*. International Pharmaceutical Federation (FIP).
Sacre, H., Hallit, S., Hajj, A., Zeenny, R.M., Akel, M., Raad, E. and Salameh, P., 2020. Developing core competencies for pharmacy graduates: The lebanese experience. *Journal of Pharmacy Practice*, p. 0897190020966195.
Warolin, C., 2003. The foundation of the school of pharmacy in Paris. *Revue D'histoire de la Pharmacie*, 51(339), pp. 453–474.

5

Teaching the Theory

5.1 Distribution of Theory Courses/ Parts Through the Curriculum

Pharmacy program curriculum is classified as courses distributed through the following departments/sciences: biomedical sciences, pharmaceutical sciences, clinical, social/administrative/behavioral sciences, and clinical sciences. Each course has credit hours and is classified as: theory (100%); mixed theory and practical or tutorial and training at Introductory Pharmacy Practice Experiences (IPPE) and Advanced Pharmacy Practice Experiences (APPE) levels. IPPE and APPE represent about 300 actual training hours (7.5 credit hours) in community and institutional health system settings and 1,440 actual training hours (36 credit hours) at hospitals in general. IPPE and APPE represented about 20% to 30% of the curriculum (ACPE, 2015), but this classification can be different from one country to another and depends on the type of pharmacy program. However, the rest of the curriculum credit hours are distributed among the following departments/sciences: biomedical sciences, pharmaceutical sciences, clinical, social/administrative/behavioral sciences, and clinical sciences. Theory parts are more during the first years and less during the advanced years as well as APPEs.

5.2 Best Practices in Teaching Online Theory

The key for success in the teaching of theory courses/parts online requires well-designed plans including the following:

Prepare the online education platforms and learning management systems and online meetings and live streaming of classes such as, Microsoft Teams (MS Teams), Google Meet, WebEx, Zoom, and other platforms or other technologies and training workshops for students and educators.

DOI: 10.1201/9781003230458-6

Design and reform the course: First, it is very important to redesign the course to be applicable for online teaching as follows:

Revise the course learning objectives and outcomes and be sure that they will be achievable with the online teaching, and that they are applicable. In general, all/majority of theory courses/parts are applicable and all related objectives and learning outcomes are achievable.

Prepare the learning outcomes for each week and link them with the course learning outcomes.

Determine the appropriate, effective teaching strategies and assessment/ evaluation methods for each topic as well as the whole course.

Prepare the online lecture notes and learning resources.

Training students and educators is very important and the key for success in online teaching.

Prepare clear guidelines and manuals for the online education platforms and learning management systems in an easy language with pictures/ screenshots from the beginning to the end.

Technology and technical support are very important for students and educators.

Adapt and implement the active/interactive teaching strategies for online teaching.

Engage and motivate students to participate effectively.

Use the technologies to enrich the teaching and learning.

Record the classes, evaluate yourself, and ask colleagues to evaluate you.

Feedback from students is very important.

Support/technical support for students and educators is very important throughout the semester.

5.3 Teaching Strategies for Online Theory

There are many effective teaching strategies that can be used to teach the theory courses/parts such as the following:

Interactive Lecture-Based Teaching Strategy

Despite the revolution in teaching strategies and the many active teaching strategies that have been developed and implemented successfully in pharmacy education around the world, lecture-based teaching strategy remains a backbone of pharmacy education and the preferred teaching strategy among pharmacy educators in many countries. However, pharmacy educators can make the online lecture interactive in many ways, such as by asking students questions every 5 to 10 minutes, presenting short videos/audios, rallying, and team sharing groups, which will attract students to the lectures. Link the theory part with life, share your practice experience with students with mini and long cases, and give students time to think about it and solve it. Engage all students and remember that many students may be hesitant to participate. Encourage all to participate. Remember that as a pharmacy educator, you are teaching the students and assessing their needs. Understanding can help also. Weekly and monthly feedback from students and colleagues can improve online teaching. Record the lectures to give to students as well as for yourself and colleagues. Feedback is the key to success in teach theory online.

Online Blended Teaching Strategy

Traditionally, blended learning combines online educational materials and opportunities for interaction online with traditional place-based classroom methods. It requires the physical presence of both teacher and student, with some elements of student control over time, place, path, or pace. However, as a result of lockdown during the COVID-19 pandemic, it was necessary to modify the face-to-face part to be online with the help of new technologies such as mobile technologies, telecommunication, and online meetings to teach theory.

Team-Based Learning (TBL) Teaching Strategy

Team-based learning (TBL) has been implemented successfully in pharmacy education teaching in many developed countries and developing countries. However, implementing and redesigning team-based learning with the help of new technologies can be an effective online teaching strategy for theory.

Problem-Based Learning (PBL) Teaching Strategy

Problem-based learning (PBL) has been implemented successfully in pharmacy education teaching in many developed countries and developing countries. However, redesigning problem-based learning with the help of new technologies such as online peer-to-peer platforms can be an effective online teaching strategy for theory.

Video-Based Learning

Short videos can be used as an effective online teaching strategy for theory.

Simulation

Role play and other simulation methods can be used with the help of new technologies as an effective online teaching strategy for theory.

Project-Based Learning

Project-based learning is a model in which the student uses online assignments and other projects, which can be used as an effective online teaching strategy for theory.

Journal Club

To critically evaluate recent articles in the academic literature, a journal club can be used as an effective online teaching strategy for theory.

Case Studies Discussion

Case studies discussion is very important and an effective online teaching strategy. Encourage students to read the given cases individually or as teams and solve them.

Self-Directed Learning

Self-directed learning allows students to improve their skills towards self-learning.

Flipped Teaching

An effective strategy that can be used online is to ask students to watch videos related to the course and to summarize them in a report.

Community Services-Based Learning

Many theory courses can use this effective teaching strategy, which allows students to achieve the course learning outcomes while contributing to patients, the public, and society.

Seminars

Seminars are an effective strategy that can be used online to improve students' presentation skills.

5.4 Assessment and Evaluation Methods for Online Theory

There are three types of assessments and evaluation: diagnostic, formative, and summative. All three types can be used for evaluation and assessment during online pharmacy education as follows:

Diagnostic assessment can help you identify your students' current knowledge, strengths, and weaknesses, which helps pharmacy educators to design, plan, and deliver the course online accordingly.

Formative assessment can measure the students' progress as well as educators' progress.

Summative assessment is used to measure the achievements of students and course learning outcomes.

5.5 Tips for Best Practices in Assessment and Evaluation

Combine the three types of assessments and evaluation: diagnostic, formative, and summative for each course when applicable.

Determine the appropriate, effective assessment/evaluation methods for each topic as well as the whole course at the beginning of the semester, discuss it in the department meeting, curriculum committee meetings, and with colleagues for validation and improvement.

Link it with the course learning objectives and outcomes as well as program learning outcomes.

Validity, reliability, and clarity are very important issues to success.

Explain it to the students at the beginning of the course.

Assess the students' diversity and needs.

Mock exams are very important for students as well as educators as they can identify the challenges in order to design an effective plan to solve them and avoid them in the future.

Feedback is very important to improve the assessment and evaluation methods.

Support/technical support for students and educators is very important throughout the semester.

5.6 Achievements

Nowadays, the majority of pharmacy schools/departments in developed countries as well as in many developing countries have adapted many active and effective teaching strategies, assessments, and evaluation methods for online teaching.

5.7 Challenges and Recommendations

Lack of technologies and resources among pharmacy schools/departments, pharmacy educators, and students are common in many low-income developing countries. Adapt and implement affordable and applicable teaching strategies and assessment and evaluation methods to overcome these challenges.

5.8 Conclusion

This chapter has discussed the teaching of theory online. This chapter includes the distribution of theory courses and parts throughout the curriculum. It also describes the teaching strategies and best practices in online teaching, assessment and evaluation methods and their best practices, and describes the achievements challenges and suggests recommendations to overcome and solve the identified challenges. The majority of pharmacy schools/departments in developed countries as well as in many developing countries have adapted many active and effective teaching strategies, assessments, and evaluation methods for online teaching.

Reference

Accreditation Council for Pharmacy Education., 2015. *Accreditation standards and key elements for the professional program in pharmacy leading to the doctor of pharmacy degree* (Standards 2016). Available at: www.acpe-accredit.org/pdf/Standards2016FINAL.pdf

6

Teaching the Practice and Tutorial

6.1. Distribution of Practicals/Laboratories and Tutorials throughout the Curriculum

Technology has played an important role in education, which has led to replacing or decreasing the need for traditional laboratories (labs) such as adapting simulation for pharmacology, anatomy, and other subjects, which lead to improving learning outcomes and, moreover, saving the pharmacy schools' budgets. However, not all pharmacy schools/departments, especially in developing countries, adapted the simulation to replace certain laboratories. The percentage of practicals/labs/tutorials throughout the pharmacy program curriculum are different from one country to another, even from one school to another. However, usually the practicals/labs/tutorials are related to the courses in the following departments/sciences: biomedical sciences, pharmaceutical sciences, clinical, social/administrative/behavioral sciences, and clinical sciences (ACPE, 2015). All practicals/labs have at least 1 credit hour, while tutorials do not. Tutorials usually elaborate on the theory part, explain more case studies, and so on. Practicals/labs are more in the biomedical sciences and pharmaceutical sciences related courses than social/administrative/behavioral sciences and clinical sciences, which have more tutorials, simulations, and case studies, and involve the student in problem solving that presents a practical application of scientific and clinical knowledge in the context of patient-centered pharmaceutical care. Teaching practical/laboratory as part of many courses throughout the pharmacy education curriculum is very important to achieving the courses' and programs' objectives and learning outcomes. However, there are many effective strategies to overcome this challenge such as: adapting simulation as a teaching strategy could be used to overcome and solve this challenge for many courses; teaching the practical/laboratory at pharmacy school/department labs in 1 week or more instead of once weekly throughout the semester and applying the local guidelines regarding the lab size and so on is another consideration.

DOI: 10.1201/9781003230458-7

6.2 Best Practices in Teaching Online Practicals/ Labs/Tutorials

The key to success in the teaching of practicals/labs/tutorials online requires well-designed plans including the following:

> Prepare the online education platforms and learning management systems and online meetings and live streaming of classes with Microsoft Teams (MS Teams), Google Meet, WebEx, Zoom, and other platforms or other technologies and training workshops for students and educators.

Design and reform the practical/labs part of courses to be applicable for online teaching as follows:

> Revise the practical/labs part of course learning objectives and outcomes and be sure that they will be achievable with online teaching, and that they are applicable.
>
> Prepare the learning outcomes for each week and link them with the course learning outcomes.
>
> Determine the appropriate, effective teaching strategies and assessment/evaluation methods for each practice as well as the whole course.
>
> Prepare the online learning resources.
>
> Training students and educators is very important and the key to success in online teaching.
>
> Prepare clear guidelines and manuals for the online education platforms and learning management systems in an easy language with pictures/screenshots from the beginning to the end.
>
> Technology and technical support are very important for the students and educators.
>
> Adapt and implement active/interactive teaching strategies for online teaching.
>
> Engage and motivate students to participate effectively.
>
> Use the technologies to enrich the teaching and learning.
>
> Record the classes, evaluate yourself, and ask colleagues to evaluate you.
>
> Feedback from students is very important.
>
> Adapt the best practices from other schools/departments in your country as well as from international experiences.

6.3 Teaching Strategies for Biomedical Sciences Related Practicals/Labs/Tutorials Online

The type of biomedical courses practicals/labs are different from one country to another as well as inside the country, and many schools have reduced the number of practical labs in the biomedical-related courses and replaced them with videos, simulation, and other methods. However, there many courses related to biomedical sciences with practicals/labs. Biomedical sciences practicals/labs are traditionally taught on-campus, with in-person sessions in a laboratory setting. In addition to the faculty members (Professors, lecturers) teaching the laboratory, teaching associates or assistants (TAs) or demonstrators facilitate the teaching of practicals/labs and support students as they complete their experiments and other practicals, and technical laboratory staff assist the educators in setting up the laboratory experiments.

Many effective teaching strategies can be used to teach the biomedical courses' related practical/labs such as following:

Virtual Laboratories

Virtual labs have been implemented successfully in many biomedical-related courses such as: Virtual physiology lab using "PhysioEx™ 10.0 Laboratory Simulations in Physiology," which provides newly formatted exercises in HTML for increased stability, web browser flexibility, and improved accessibility. The 12 exercises contain 63 easy-to-use laboratory simulation activities that complement or replace wet labs, including those that are expensive or time-consuming to perform in class (Peter et al., 2020).

Practicals/Labs Recording

Educators can record the practicals/labs and send them to students, as well as demonstrate it online for students using online meetings.

Project-Based Learning

Project-based learning is a model in which the student uses online assignments and other projects, which can be used as an effective online teaching strategy for the practicals/labs.

Virtual Resources

Virtual resources play an important role in the success of delivering the practicals/labs for many courses such as: virtual cadaveric resources such as Acland's Video Atlas of Human Anatomy (Lippincott Williams & Wilkins,

PA) and 3D modeling programs such as Visible Body (Argosy Publishing Inc., MA) and synchronous webinars, using Zoom (Zoom Voice Communications Inc., CA), or Microsoft Teams (Microsoft, WA).

6.4 Teaching Strategies for Pharmaceutical Sciences Related Practicals/Labs/Tutorials Online

Pharmaceutical sciences practicals/labs are traditionally taught on-campus, with in-person sessions in a laboratory setting. In addition to the faculty members (Professors, lecturers) teaching the laboratory, teaching associates or assistants (TAs) or demonstrators facilitate the teaching of practicals/labs and support students as they complete their experiments and other practicals and technical laboratory staff assist the educators in setting up the laboratory experiments.

Many effective teaching strategies can be used to teach the pharmaceutical courses related practicals/labs such as the following:

Virtual Laboratories

Virtual labs have been implemented successfully in many pharmaceutical-related courses such as:

Virtual pharmacology using (CyberPatient), an organ bath simulator, (OBSim), AutonomiCAL, Virtual Cat, and RatCVS (Ezeala et al., 2020).

Practicals/Labs Recording

Educators can record the practicals/labs and send them to students, as well as demonstrate it online for students using online meetings.

Project-Based Learning

Project-based learning is a model in which the student works online for assignments and other projects and can be used as an effective online teaching strategy for the practicals/labs.

Virtual Resources

Virtual resources play an important role in the success of delivering the practicals/labs for many courses such as: virtual cadaveric resources.

6.5 Teaching Strategies for Social/Administrative/ Behavioral and Clinical Sciences Related Practicals/Labs/Tutorials Online

Tutorials are more common in the social/administrative/behavioral sciences than in the clinical sciences. However, practicals/labs are also common practice in many pharmacy schools/departments for many courses. Many effective teaching strategies can be used to teach the social/administrative/ behavioral sciences and clinical sciences related practicals/labs such as the following:

Simulation/Virtual Learning

Simulation/virtual learning have been implemented successfully in many social/administrative/behavioral sciences and clinical sciences; for example, computer/mobile softwares for clinical pharmacokinetics/therapeutic drug monitoring (TDM) such as bayesian software and others. MyDispense (by Monash University) allows students to develop and practice their dispensing skills.

Role Play

Role play can help as an effective strategy for many social/administrative/ behavioral sciences and clinical sciences related practicals/labs and tutorials.

Practicals/Labs and Tutorials Recording

Educators can record the practicals/labs or tutorials and send them to students, as well as demonstrate it online for students using online meetings.

Project-Based Learning

Project-based learning is a model in which the student works online for assignments and other projects and can be used as an effective online teaching strategy for the practicals/labs.

Virtual Resources

Virtual resources play an important role in the success of delivering the practicals/labs for many courses.

Journal Club

To critically evaluate recent articles in the academic literature, journal clubs can be used as an effective online teaching strategy.

6.6 Assessment and Evaluation Methods for Online Practicals/Labs

There are three types of assessments and evaluation: diagnostic, formative, and summative. All three types can be used for evaluation and assessment during online pharmacy education as follows:

Diagnostic assessment can help you identify your students' current knowledge, strengths, and weaknesses which help pharmacy educators to design, plan, and deliver the course online accordingly.

Formative assessment can measure the students' progress as well as educators' progress.

Summative assessment can measure the achievements of students and course learning outcomes.

6.7 Tips for Best Practices in Assessment and Evaluation

Combine the three types of assessments and evaluation: diagnostic, formative, and summative for each course when applicable.

Determine the appropriate, effective assessment/evaluation methods for each practical/lab as well as the whole practical/lab at the beginning of the semester, discuss it in the department meeting, curriculum committee meetings, and with colleagues for validation and improvement.

Link it with the course learning objectives and outcomes as well as program learning outcomes.

Explain it to the students at the beginning of the course.

Assess the students' diversity and needs.

Identify the challenges and design an effective plan to solve them and avoid them in the future.

Feedback is very important to improve the assessment and evaluation methods.

Objective Structured Clinical Examination (OSCE) can be used effectively online to evaluate students' competencies.

6.8 Achievements

Nowadays, the majority of pharmacy schools/departments in developed countries as well as in many developing countries have adapted many active and effective teaching strategies, assessments, and evaluation methods for online practicals/labs and tutorials.

6.9 Challenges and Recommendations

Lack of technologies and resources among pharmacy schools/departments, pharmacy educators, and students are common in many low-income developing countries. Adapt and implement the affordable and applicable teaching strategies and assessment and evaluation methods to overcome these challenges.

6.10 Conclusion

This chapter has discussed teaching practicals/labs and tutorials online. This chapter includes the teaching strategies and best practices in online teaching, assessment and evaluation methods and its best practices, describes the achievements and challenges and suggests recommendations to overcome and solve the identified challenges. The majority of pharmacy schools/departments in developed countries as well as in many developing countries have adapted many active and effective teaching strategies, assessments, and evaluation methods for online practicals/labs and tutorials.

References

Accreditation Council for Pharmacy Education., 2015. *Accreditation standards and key elements for the professional program in pharmacy leading to the doctor of pharmacy degree* (Standards 2016). Available at: www.acpe-accredit.org/pdf/Standards2016FINAL.pdf

Ezeala, C.C., Ezeala, M.O. and Akapelwa, T.M., 2020. A survey of medical students'
    experiences with online practical pharmacology classes during Covid19 lock-
    down. *Medical Journal of Zambia*, 48(1), pp. 25–30.
Peter, Z., Timothy, N., Stabler, Lori A., Smith, Andrew Lokuta and Griff, E. 2020.
    PhysioEx 10.0: Laboratory simulations in physiology, 1st edition. Pearson.

7

Introductory Pharmacy Practice Experiences (IPPE) and Advanced Pharmacy Practice Experiences (APPE)

7.1 Distribution of Introductory Pharmacy Practice Experiences (IPPE) and Advanced Pharmacy Practice Experiences (APPE) Training through the Curriculum

Introductory Pharmacy Practice Experiences (IPPE) and Advanced Pharmacy Practice Experiences (APPE) training are is an important part of the pharmacy curriculum to prepare students to practice, to be good pharmacists able to provide good pharmacist care and patient care services. IPPE and APPE represent about 300 actual training hours (7.5 credit hours) in community and institutional health-system settings and 1,440 actual training hours (36 credit hours) at hospitals in general. IPPE and APPE represent about 20% to 30% of the curriculum (ACPE, 2015), although this classification can be different from one country to another and depends on the type of pharmacy program.

7.2 Introductory Pharmacy Practice Experiences (IPPEs)

Introductory Pharmacy Practice Experiences (IPPEs) occur in the early years of the pharmacy curriculum which allows students to begin to engage in professional pharmacy work. IPPEs provide the student pharmacist with an opportunity to observe, learn from, and work with pharmacists in a variety of practice settings and to participate in patient care activities under the supervision of preceptors. Introductory Pharmacy Practice Experiences (IPPEs) clerkships/rotations examples include:

Community Pharmacy Training
Institutional Pharmacy Introductory Pharmacy Practice Experience

DOI: 10.1201/9781003230458-8

7.3. Advanced Pharmacy Practice Experiences (APPEs)

Advanced Pharmacy Practice Experiences (APPEs) are experiential courses designed to allow student pharmacists to gain experience, apply knowledge and skills, and gain professional competence and confidence by delivering pharmacist care and patient care services under the supervision of preceptors.

7.4 Best practices in Online Introductory Pharmacy Practice Experiences (IPPEs) and Advanced Pharmacy Practice Experiences (APPEs)

The key to success in the online training requires a well-designed plan that includes the following:

Prepare the online education platforms and learning management systems and online meetings and live streaming of classes such as, Microsoft Teams (MS Teams), Google Meet, WebEx, Zoom, and other platforms or technologies and training workshops for students and educators.

Design and reform the training-related courses to be applicable for online teaching and learning as follows:

Revise the training learning objectives and outcomes and be sure that they will be achievable with online teaching and learning, and are applicable.

Prepare the learning outcomes for each week and link them with the course learning outcomes.

Determine appropriate, effective teaching strategies and assessment/ evaluation methods.

Prepare the online learning resources.

Training students and educators is very important and the key to success in online teaching.

Prepare clear guidelines and manuals for the online education platforms and learning management systems in an easy language with pictures/screenshots from the beginning to the end.

Technology and technical support are very important for the students and educators.

Adapt and implement the active/interactive teaching strategies for online teaching.

Engage and motivate students to participate effectively.

Use the technologies to enrich the teaching and learning.

Feedback from students is very important.

Adapt the best practices from other schools/departments in your country as well as from the international experiences.

Simulation-Based Training

Simulation-based training can be used effectively to train students for IPPEs and APPEs as follows:

Simulation Cases

Pharmacy educators can develop and adapt case studies, and provide them to students to solve the cases, develop management plans for the patients, identify the drug-related problems (DRPs) and solve or prevent them.

Simulation Prescriptions and Orders

Pharmacy educators can develop and adapt medical prescriptions and orders and provide them to students to identify the prescription writing errors.

Simulation/Virtual Learning

This includes computer and mobile software for clinical pharmacokinetics/ therapeutic drug monitoring (TDM) such as bayesian software and others. MyDispense (by Monash University) allows students to develop and practice their dispensing skills.

Role Play

Role play can be used in IPPEs and APPEs for pharmacist care/patient care services, patients' interview, and patient education and counseling.

Simulation Consultation Services

This involves using technologies and mobile technologies for effective online simulation consultation services for APPEs.

Simulation Drug Information Services

This involves using technologies and mobile technologies for the effective online simulation of drug information services.

Simulation Pharmacovigilance and Adverse Drug Reactions (ADRs) Reporting

This involves using technologies and mobile technologies for the effective online simulation of Adverse Drug Reactions (ADRs) reporting.

Simulation Medication Errors Reporting

This involves using technologies and mobile technologies for the effective online simulation of medication errors reporting.

Project-Based Learning

Project-based learning is a model in which the student works online for assignments and other projects, which can be used as an effective online training strategy.

Virtual Resources

Virtual resources play an important role in the success of online training strategy.

Journal Club

To critically evaluate recent articles in the academic literature, a journal club can be used as an effective online training strategy.

7.5 Tips for Best Practices in Assessment and Evaluation

Combine the three types of assessments and evaluation: diagnostic, formative, and summative for each course when applicable.

Determine the appropriate, effective assessment/evaluation methods for each clerkship/training as well as the whole practical/labs at the beginning of the semester, discuss it in the department meeting, curriculum committee meetings, and with colleagues for validation and improvement.

Link them with the course learning objectives and outcomes as well as program learning outcomes.

Explain them to students at the beginning of the course.

Assess the students' diversity and needs.

Identify the challenges and design an effective plan to solve them and avoid them in the future.

Feedback is very important to improve the assessment and evaluation methods.

Objective Structured Clinical Examination (OSCE) can be used effectively online to evaluate students' competencies.

7.6 Achievements

Nowadays, the majority of pharmacy schools/departments in developed countries as well as in many developing countries have adapted many active and effective teaching strategies, assessments, and evaluation methods for online training.

7.7 Challenges and Recommendations

Lack of technologies and resources among pharmacy schools/departments, pharmacy educators, and students are common in many low-income developing countries. Adapting and implementing affordable training can overcome these challenges.

7.8 Conclusion

This chapter has discussed online Introductory Pharmacy Practice Experiences (IPPE) and Advanced Pharmacy Practice Experiences (APPE) training. This chapter includes the best practices in online Introductory Pharmacy Practice Experiences (IPPE) and Advanced Pharmacy Practice Experiences (APPE) training, describes the achievements and challenges, and suggests recommendations to overcome and solve the identified challenges. Lack of technologies and resources among pharmacy schools/departments, pharmacy educators, and students are common in many low-income developing countries. Adapting and implementing affordable training can overcome these challenges.

Reference

Accreditation Council for Pharmacy Education., 2015. *Accreditation standards and key elements for the professional program in pharmacy leading to the Doctor of Pharmacy degree* (Standards 2016). Available at: https://www.acpe-accredit.org/pdf/Standards2016FINAL.pdf

8

Technologies and Tools

8.1 Technologies and Tools in Online Pharmacy Education

Suitable and effective technologies and tools are the keys to success in online pharmacy education. Pharmacy schools/departments should use and adapt the most effective technologies and tools for teaching and learning, and prepare the educators and students with all required technologies as follows:

Internet

The internet plays a very important and vital role in online pharmacy education. The internet facilitates the teaching and learning process, which makes distance/online pharmacy education more effective and easier than at any time in history. Pharmacy educators and students need the internet to: make communication easy; deliver the classes; upload/download the lecture notes and other educational materials/resources; exams; assignments; presentations; search for pharmacy-related information; train; explore patient care services; promote public health, awareness, and services. Without access to the internet, online teaching and learning will be impossible.

Computers and Laptops

Computers and laptops are very important and essential for online teaching and learning. They provide flexible and effective access to online teaching and learning for pharmacy educators and students.

Smartphones, Tablets, and Net Books

Using smartphones, tablets, and net books is very important and essential for online teaching and learning. They provide flexible and effective access to online teaching and learning for pharmacy educators and students. They help students to download many education resources. Furthermore, students can access them at any time throughout the day inside or outside the home.

DOI: 10.1201/9781003230458-9

It helps educators also to prepare, revise, and access lecture notes, literature, educational resources, and others at any time and any place.

Learning Management Systems (LMS)

Learning Management Systems (LMS) are very important in online pharmacy education as well as higher education, as they contain an effective web-based learning system of sharing study materials, making announcements, conducting evaluation and assessments, generating results, and communicating interactively in synchronous and asynchronous ways among various other academic activities (Kant et al., 2021; Bervell and Umar, 2017).

Moodle

Moodle is an open-source LMS that provides collaborative learning environments which empower learning and teaching. It is a flexible and user-friendly platform adopted by most educational institutions and businesses of all sizes.

Blackboard

Blackboard is the most popular LMS used by businesses and educational institutes worldwide, which delivers a powerful learning experience. It is easily customizable according to your organization's needs. It provides advanced features and integrates with Dropbox, Microsoft OneDrive, and school information systems.

Moodle and Blackboard are two of the most famous and widely known Learning Management Systems (LMS) in pharmacy education and help educators as well students in the teaching and learning process (Momani, 2010).

Momani (2010) compares the two known Learning Management Systems (LMS) in terms of Pedagogical Factor, Learner Environment, Instructor Tools, Course and Curriculum Design, Administrator Tools, and technical specifications and report that they have lots in common, but also have some key differences which make each one special in its own way (Momani, 2010).

Webinar and Video Conferencing Platforms

Webinar and video conferencing platforms are very important in online pharmacy education. Microsoft Teams, Cisco WebEx Teams, Google Meet, and Zoom are the most common Webinar and Video Conferencing Platforms in online pharmacy education (www.microsoft.com/en-us/microsoft-teams/group-chat-software; www.webex.com/; https://apps.google.com/meet/; www.zoom.us/).

Microsoft Teams, Cisco WebEx Teams, Google Meet, and Zoom offer multiple versions of their software based on usage requirements. This includes free versions that are great for light uses, short conference calls, and light file sharing. However, universities pay the cost of all platforms for pharmacy educators and students to facilitate teaching and learning. All platforms are used successfully in online pharmacy education worldwide. Training, workshops, and writing manuals are very important for both pharmacy educators and students in order to use them successfully.

YouTube

YouTube can be used for streaming to the audience and having a one-way video interaction. YouTube can be a good resource for many pharmacy-related education videos. Pharmacy educators can use YouTube for recording lectures and sharing them with students as pharmacy students can easily access it at any time, download it if needed, as well as share it with their colleagues.

Facebook, Twitter, and Instagram

Facebook, Twitter, and Instagram can be used in online pharmacy education in many ways as follows: allows students to interact with their mentors, access their course contents, customising and building student communities; transferring resource materials, collaborative learning, and interaction with colleagues as well as teachers, which would facilitate students being more enthusiastic and dynamic; sharing health information, discussing clinical cases, and listening to patient stories (Wong et al., 2019; Shafer et al., 2018; Gulati et al., 2020)

WhatsApp

WhatsApp can be used in online pharmacy education teaching and learning in many ways as follows: communication tools between the pharmacy educators and students and between preceptors and students in training; share lecture notes and other educational resources; group chat and others (Coleman and O'Connor, 2019; Raiman et al., 2017; us Salam et al., 2021).

Wearable Technologies

Wearable technologies have a tremendous potential to improve education, empowering students as well as instructors in their teaching and learning experiences (Borthwick et al., 2015; Subrahmanyam and Swathi, 2020). Wearable technologies can be used in online pharmacy education for the following potential reasons: improved student engagement; convenient to wear as they are hands-free gadgets; effortless communication with enhanced features; facility to record videos; teaching through augmented reality; teaching through virtual reality and wearables–learning apps (Subrahmanyam and Swathi, 2020).

8.2 Online and Digital Library

An online and digital library is a collection of documents such as journal articles, books, and other educational resources organized in an electronic form and available on the internet for students, educators, and other staff. Furthermore, access to the databases helps students, educators, and staff to access the latest volumes/issues of scientific journals.

8.3 Achievements

Nowadays, the majority of pharmacy schools/departments in developed countries as well as in many developing countries have adapted many effective technologies strategies, assessments, and evaluation methods for online training.

8.4 Challenges and Recommendations

Lack of technologies and resources among pharmacy schools/departments, pharmacy educators, and students is common in many low-income developing countries. Adapting and implementing affordable training can overcome these challenges.

8.5 Conclusion

This chapter has discussed the technologies and resources for effective online teaching and learning. Suitable and effective technologies and tools are the keys to success in online pharmacy education. Pharmacy schools/departments should use and adapt the most effective technologies and tools for teaching and learning and prepare the educators and students with all required technologies.

References

Bervell, B. and Umar, I.N., 2017. A decade of LMS acceptance and adoption research in Sub-Sahara African higher education: A systematic review of models, methodologies, milestones and main challenges. *EURASIA Journal of Mathematics, Science and Technology Education*, 13(11), pp. 7269–7286.

Borthwick, A.C., Anderson, C.L., Finsness, E.S. and Foulger, T.S., 2015. Special article personal wearable technologies in education: Value or villain? *Journal of Digital Learning in Teacher Education*, 31(3), pp. 85–92.

Coleman, E. and O'Connor, E., 2019. The role of WhatsApp® in medical education: A scoping review and instructional design model. *BMC Medical Education*, 19(1), pp. 1–13.

Gulati, R.R., Reid, H. and Gill, M., 2020. Instagram for peer teaching: Opportunity and challenge. *Education for Primary Care*, 31(6), pp. 382–384.

Kant, N., Prasad, K.D. and Anjali, K., 2021. Selecting an appropriate learning management system in open and distance learning: A strategic approach. *Asian Association of Open Universities Journal*, 16(1), pp. 79–97.

Momani, A.M., 2010. Comparison between two learning management systems: Moodle and blackboard. *Behavioral & Social Methods eJournal*, 2(54).

Raiman, L., Antbring, R. and Mahmood, A., 2017. WhatsApp messenger as a tool to supplement medical education for medical students on clinical attachment. *BMC Medical Education*, 17(1), pp. 1–9.

Shafer, S., Johnson, M.B., Thomas, R.B., Johnson, P.T. and Fishman, E.K., 2018. Instagram as a vehicle for education: What radiology educators need to know. *Academic Radiology*, 25(6), pp. 819–822.

Subrahmanyam, V.V. and Swathi, K., 2020. *Wearable technology and its role in education*. International Conference—2020 on Distance Education and Educational Technology (ICE-CODL 2020), CDOL, JMI, New Delhi 10–11 December 2020.

us Salam, M.A., Oyekwe, G.C., Ghani, S.A. and Choudhury, R.I., 2021. How can WhatsApp® facilitate the future of medical education and clinical practice? *BMC Medical Education*, 21(1), pp. 1–4.

Wong, X.L., Liu, R.C. and Sebaratnam, D.F., 2019. Evolving role of Instagram in# medicine. *Internal Medicine Journal*, 49(10), pp. 1329–1332.

9

Self-Learning and Self-Directed Learning

9.1 Background

Self-learning or self-directed learning was defined by Knowles as a process in which a learner takes the initiative, diagnoses their learning needs, creates learning goals, identifies resources for learning, applies appropriate learning strategies, and evaluates their learning outcomes (Knowles, 1975). The Accreditation Council for Pharmacy Education (ACPE) defines CPD as "the lifelong process of active participation in learning activities that assists individuals in developing and maintaining continuing competence, enhancing their professional practice, and supporting achievement of their career goals" (ACPE, 2015). The history of self-learning or self-directed learning goes back to the 1800s. In 1840 in the United States, Craik documented and celebrated the self-education efforts of several people showing early scholarly efforts to understand self-directed learning. In 1895 in Great Britain, Smiles published a book entitled *Self-Help*, that applauded the value of personal development (SDL Timeline, 2021). Pharmacy schools/departments should provide an environment and culture that promotes self-directed lifelong learning among pharmacy students to participate in self-learning/self-directed learning activities. Pharmacy schools/departments should ensure that the pharmacy curriculum includes self-learning/self-directed learning experiences and time for independent study to allow pharmacy students to develop and improve the required skills of lifelong learning. Preparing pharmacy students to be lifelong learners is very important and should be included in all pharmacy programs' competencies and learning outcomes.

9.2 Rationality of Self-Learning/Self-Directed Learning

Pharmacy practice has changed during the last decades around the world and continues changing as a result of the advances in medicine, health care and patient care practices, new technologies and systems, and process improvements, as well as changes in professional roles and responsibilities, which require pharmacists and future pharmacists to be ready for any change, and to have the required knowledge and skills to provide the most effective pharmacist care/patient care services.

9.3 Best Practices in Self-Learning/Self-Directed Learning

Best practices in self-learning/self-directed learning require good planning, goals, activities, facilities, and monitoring. The following steps can be used as a guide (Tofade et al., 2011; Briceland and Hamilto, 2010; *Self-Directed Learning (SDL)*, 2021):

Assess Readiness of Students for Self-Learning/Self-Directed Learning

Assessing the readiness of students for self-learning/self-directed learning is very important. It could be done by evaluating students' abilities, identifying their strengths and weaknesses, and making a plan to work on the weaknesses. Remember that there is a difference between students' abilities and readiness for self-learning/self-directed learning; therefore, don't expect that all students will be at the same level. Identifying the needs of students and differences is very important in their success.

Prepare Students for Self-Learning/Self-Directed Learning

Good preparations are necessary to make it possible for students to work independently and achieve excellence in self-learning/self-directed learning.

Design/Develop the Goals of Self-Learning/Self-Directed Learning

Designing/developing the goals of self-learning/self-directed learning is very important and this could be different from one school to another, from one pharmacy program to another. Schools/departments and programs should design applicable goals based on the available facilities and resources and students' needs and abilities.

Design/Develop the Objectives and Learning Outcomes

Designing the action plan with specific, measurable, achievable, relevant, and time-bound (SMART) learning objectives and outcomes is very important and should be based on the available facilities and resources and students' needs and abilities.

Engage in the Learning Process

Self-Directed Learning: A Four-Step Process, Centre for Teaching Excellence, University of Waterloo stated: Students need to understand themselves as learners in order to understand their needs as self-directed learning students—referring students to our resources on learning preferences may be helpful. Students should also consider answering the following questions:

What are my needs re: instructional methods?

Who was my favorite teacher? Why?

What did they do that was different from other teachers?

Students should reflect on these questions throughout their program and substitute "teacher" with "advising instructor."

Students also need to understand their approach to studying:

A deep approach to studying involves transformation and is ideal for self-directed learning. This approach is about understanding ideas for yourself, applying knowledge to new situations, and using novel examples to explain a concept, and learning more than is required for unit completion.

A surface approach involves reproduction: coping with unit requirements, learning only what is required to complete a unit in good standing, and tending to regurgitate examples and explanations used in readings.

A strategic approach involves organization: achieving the highest possible grades, learning what is required to pass exams, memorizing facts, and spending time practicing from past exams.

Earlier academic work may have encouraged a surface or strategic approach to studying. These approaches will not be sufficient (or even appropriate) for successful independent study. Independent study requires a deep approach to studying, in which students must understand ideas and be able to apply knowledge to new situations. Students need to generate their own connections and be their own motivators (*Self-Directed Learning (SDL)*, 2021).

Evaluate Learning

Self-Directed Learning: A Four-Step Process, Centre for Teaching Excellence, University of Waterloo reported: For students to be successful in self-directed learning, they must be able to engage in self-reflection and self-evaluation of their learning goals and progress in a unit of study. To support this self-evaluation process, they should:

- regularly consult with the advising instructor,
- seek feedback, and
- engage in reflection of their achievements, which involves asking:
  - How do I know I've learned?
  - Am I flexible in adapting and applying knowledge?
  - Do I have confidence in explaining material?
  - When do I know I've learned enough?
  - When is it time for self-reflection and when is it time for consultation with the advising faculty member? *Self-Directed Learning (SDL), 2021)*

Responsibilities of Pharmacy Schools/Departments in Self-Learning/Self-Directed Learning

Pharmacy schools/departments play a very important role in the success of self-learning/self-directed learning. Pharmacy schools/departments offer and provide prepared facilities, resources, and online resources for pharmacy educators and students; conduct training workshops and courses for pharmacy educators and students; and provide technical support to pharmacy educators and students.

Responsibilities and Roles of Pharmacy Educators in Self-Learning/Self-Directed Learning

Pharmacy educators play a very important role in the success of self-learning/self-directed learning as follows:

- Identify the students' abilities and readiness for self-learning/self-directed learning.
- Prepare students for self-learning/self-directed learning.
- Build a co-operative learning environment.
- Help to motivate and direct the students' learning experience.
- Facilitate students' initiatives for learning.
- Be available for consultations as appropriate during the learning process.
- Serve as an advisor rather than a formal instructor and educator.

Responsibilities of Pharmacy Students in Self-Learning/Self-Directed Learning

Self-assess your readiness to learn.

Define your learning goals and develop a learning contract.

Monitor your learning process.

Take initiative for all stages of the learning process—be self-motivated.

Re-evaluate and alter goals as required during your unit of study.

Consult with your advising instructor as required.

9.4 Self-Learning/Self-Directed Learning Teaching Strategies

9.4.1 Team-Based Learning (TBL) Teaching Strategy

Team-based learning (TBL) has been implemented successfully in pharmacy education teaching in many developed countries and developing countries. However, redesigning team-based learning with the help of new technologies can be an effective online teaching strategy for theory.

9.4.2 Problem-Based Learning (PBL) Teaching Strategy

Problem-based learning (PBL) has been implemented successfully in pharmacy education teaching in many developed countries and developing countries. However, redesigning problem-based learning with the help of new technologies such as online peer-to-peer platforms can be an effective online teaching strategy for theory.

9.4.3 Project-Based Learning

Project-based learning is a model in which the student works online for assignments and other projects and can be used as an effective online teaching strategy for theory.

9.4.4 Journal Club

To critically evaluate recent articles in the academic literature, a journal club can be used as an effective online teaching strategy for theory.

9.4.5 Case Studies Discussion

Case studies discussion is very important and is an effective online teaching strategy. Encourage students to read the given cases individually or as teams and solve them.

9.4.6 Flipped Teaching

Flipped teaching is an effective strategy and can be used online. For example, ask students to watch videos related to the course and summarize them in a report.

9.4.7 Community Services-Based Learning

Many theory courses can use this effective teaching strategy, which allows students to achieve the course learning outcomes while contributing to patients, the public, and society.

9.5 Conclusion

This chapter covers Self-Learning and Self-Directed Learning related issues. The Accreditation Council for Pharmacy Education (ACPE) defines CPD as "the lifelong process of active participation in learning activities that assists individuals in developing and maintaining continuing competence, enhancing their professional practice, and supporting achievement of their career goals."

References

Accreditation Council for Pharmacy Education., 2015. *Accreditation standards and key elements for the professional program in pharmacy leading to the Doctor of Pharmacy degree* (Standards 2016). Available at: https://www.acpe-accredit.org/pdf/Standards2016FINAL.pdf

Briceland, L.L. and Hamilton, R.A., 2010. Electronic reflective student portfolios to demonstrate achievement of ability-based outcomes during advanced pharmacy practice experiences. *American Journal of Pharmaceutical Education*, 74(5).

Knowles, M.S., 1975. *Self-directed learning*. Prentice Hall Regents.

SDL Timeline., 2021. *Historical timeline for self-directed learning in adult education*. Available at: http://sdlearning.pbworks.com/w/page/1899125/SDL%20Timeline

Self-Directed Learning (SDL), 2021. *A four-step process. Centre for teaching excellence, University of Waterloo*. Available at: https://uwaterloo.ca/centre-for-teaching-excellence/teaching-resources/teaching-tips/tips-students/self-directed-learning/self-directed-learning-four-step-process

Tofade, T., Franklin, B., Noell, B. and Leadon, K., 2011. *Evaluation of a continuing professional development program for first year student pharmacists undergoing an introductory pharmacy practice experience*. Available at: https://conservancy.umn.edu/handle/11299/109694

10

Continuous Pharmacy Education and Professional Development for Pharmacy Educators

10.1 Background

Pharmacy educators need continuous pharmacy education, continuous education, and professional development nowadays more than at any time in history due to many reasons such as development of teaching and learning, development of technology, and development in pharmacy practice. Pharmacy educators in general are pharmacists. Pharmacy practice has changed during the last decades around the world and continues changing as a result of advances in the medicine, health care and patient care practices, new technologies and systems, and process improvements, as well as changes in professional roles and responsibilities. Pharmacists nowadays play/should play an important role in patient care, and they contribute effectively to patients' health, diseases/conditions management as well as prevention; treating outcomes; treating cost; patients' quality of life; and satisfaction towards the health care system and care. Pharmacists nowadays provide online pharmacist care services in many countries around the world. Therefore, pharmacy educators should update and improve their knowledge and skills related to pharmacy practice and education.

10.2 Importance of Continuous Pharmacy Education and Continuous Education, Professional Development for Pharmacy Educators

10.2.1 Learn New Teaching Strategies

Learning new teaching strategies will help pharmacy educators to apply them and could be helpful in the achievement of course learning outcomes and program learning outcomes.

DOI: 10.1201/9781003230458-11

10.2.2 Learn New Assessment and Evaluation Methods

Learning new assessment and evaluation methods will help pharmacy educators to apply them and to provide better assessment and evaluation for the course and the program.

10.2.3 Develop Better Organization and Planning Skills

Continuous education and professional development training can help pharmacy educators to plan and manage their time and work.

10.2.4 Improve and Update Knowledge

Many online continuous education and professional development programs are available and accessible. They can enable pharmacy educators to expand, improve, and update their knowledge in different areas.

10.2.5 Improve and Update Skills

Many online continuous education and professional development programs are available and accessible. They can enable pharmacy educators to expand, improve, and update their skills.

10.2.6 Leadership

Many online continuous education and professional development programs focus on leadership skills.

10.2.7 Communication

Many online continuous education and professional development programs focus on communication skills. Improving communication skills, learning new techniques and skills can help pharmacy educators to communicate effectively with their students, patients, the public, and health care professionals. Good and effective communication is the key to success.

10.2.8 Technology

Many online continuous education and professional development programs focus on technology, and its impact on education and practice. Improving technology-related knowledge and skills is very important for pharmacy educators.

10.2.9 Design the Course and Deliver It Online

Many online continuous education and professional development programs focus on how to design the course and deliver it online.

10.2.10 Learning Management Systems (LMS)

Many online continuous education and professional development programs focus on the Learning Management Systems (LMS) and how you can use them effectively, which is very important for pharmacy educators in online pharmacy education.

10.2.11 Webinar and Video Conferencing Platforms

Many online continuous education and professional development programs focus on webinar and video conferencing platforms, which are very important for pharmacy educators in online pharmacy education.

10.2.12 Online and Digital Library

Many online continuous education and professional development programs focus on online and digital libraries and how you can use them effectively in online education, which is very important for pharmacy educators in online pharmacy education.

10.2.13 YouTube

Many online continuous education and professional development programs focus on YouTube and how you can use it effectively in online education. YouTube can be used for streaming to the audience and having a one-way video interaction. YouTube can be a good resource for pharmacy-related education videos. Pharmacy educators can use YouTube for recording lectures and sharing them with students. On YouTube pharmacy students can easily access lectures at any time, download them if needed, as well as share them with their colleagues, which is very important for pharmacy educators in online pharmacy education.

10.2.14 Social Media

Many online continuous education and professional development programs focus on social media and how you can use it effectively in online education. Social media can be used in online pharmacy education teaching and learning in many ways as follows: communication tools between pharmacy educators and students and preceptors and students in training; sharing lecture notes and other educational resources; group chats and others, which is very important for pharmacy educators in online pharmacy education.

10.2.15 Electronic Files

Attending online continuous education and professional development programs about electronic files is very important for pharmacy educators in online pharmacy education.

10.2.16 Academic Advising

Attending online continuous education and professional development programs about academic advising is very important for pharmacy educators in online pharmacy education.

10.2.17 Electronic Portfolio

Maintaining a professional electronic portfolio is very important for pharmacy educators and students in online pharmacy education.

10.2.18 Online and Distance Patient Care

Attending online continuous education and professional development programs about online patient care services is very important for pharmacy educators with interest in patient care.

10.3 Best Practices in Continuous Pharmacy Education and Continuous Education, Professional Development for Pharmacy Educators

Best practices in continuous education and professional development require good plans, goals, activities, facilities, and monitoring (Tofade et al., 2011; Briceland and Hamilto, 2010) as follows:

Identifying Your Needs

Identifying your needs for continuous education and professional development is very important and this is the cornerstone of continuous education and professional development for pharmacy educators.

Designing/Developing the Goals and Objectives of Continuous Education and Professional Development

Designing/developing the goals of continuous education and professional development is very important and this could be different from one pharmacy educator to another.

Identifying Activities, Courses, Programs, Workshops, Conferences, and Others

Identifying activities, courses, programs, workshops, conferences, and other activities for continuous education and professional development is very important for pharmacy educators' development.

Applying the Self-Learning Process Approach

Pharmacy educators by nature are life-long learners. Applying the concept/approach of self-learning is very important for your continuous education and professional development.

Reflecting on Your Learning

Reflecting on what you have learned is a vital part of continuous education and professional development.

Applying Your Learning to Your Teaching and Practice

Applying what you have learned to your teaching and practice is the best way to practice it and learn from your practice.

Sharing Learning with Your Colleagues

Sharing what you have learned with your colleagues is very important for you as you will practice the ability of teaching others and learn from it.

Documentation

Documentation is very important in continuous education and professional development.

10.4 Conclusion

This chapter has discussed continuous pharmacy education and professional development for pharmacy educators related issues. Pharmacy educators in general are pharmacists. Pharmacy practice has changed during the last decades around the world and continues changing as a result of advances in medicine, health care, and patient care practices; new technologies, systems, and process improvements; as well as changes in professional roles and responsibilities. Pharmacy educators need continuous pharmacy education, continuous education, and professional development nowadays more than at any time in history due to many reasons such as development of teaching and learning, development of technology, and development in pharmacy practice.

References

Briceland, L.L. and Hamilton, R.A., 2010. Electronic reflective student portfolios to demonstrate achievement of ability-based outcomes during advanced pharmacy practice experiences. *American Journal of Pharmaceutical Education*, 74(5).

Tofade, T., Franklin, B., Noell, B. and Leadon, K., 2011. *Evaluation of a continuing professional development program for first year student pharmacists undergoing an introductory pharmacy practice experience.* Available at: https://conservancy.umn.edu/handle/11299/109694

11

Community Services

11.1 Background

Pharmacy students are the future pharmacists, and they will play an important role in educating the public about their diseases, conditions, and health, and increasing their awareness about their medications, the importance of adherence towards non-pharmacological interventions and pharmacological interventions, medication safety issues, risk factors for developing diseases and how to avoid them, how to modify the modifiable risk factors, and awareness of risk factors and diseases screening, which allow early interventions to modify the modifiable risk factors such as educating potential patients to quit smoking, stop drinking alcohol, eat a healthy diet, get involved in exercise and other interventions. Furthermore, it is important to help patients to control blood pressure, cholesterol, and blood glucose by non-pharmacological interventions or pharmacological interventions if needed, which will contribute effectively to preventing angina. Patients with diabetes mellitus, hypertension, and hyperlipidemia should go for cardiovascular risk factors screening. Community services by pharmacy schools/departments play an important role in the health care system and have a good impact on patients' life and health, improving treatment outcomes, improving the quality of life, improving satisfaction, and helping the ministries of health in patient care services.

11.2 Importance of Community Services to Pharmacy Students

Pharmacy schools/departments aim to graduate students with the necessary competencies that help them to be good pharmacists and contribute effectively to patients and the public and to provide effective pharmacist care/patient care services. One of the most important competencies is related to community services and public education and health promotion (ACPE, 2015). Volunteering in community services allows pharmacy students to give

DOI: 10.1201/9781003230458-12

back to their communities, develop skills needed for practice in an ambulatory care or community setting, gain insight into the lives of people who are underserved, and learn about people of different cultures and backgrounds (Farzadeh, 2019). Faradei reports the following benefits:

Developing Empathy

Empathy involves acknowledging patients' concerns, incorporating patients' perspectives into care, and demonstrating appropriate verbal and non-verbal language. Empathy is essential in developing successful relationships between pharmacists and patients and optimizing clinical outcomes. Educating the underserved can help pharmacy students develop empathy for their patients.

Developing Communication Skills for Various Age Groups

Pharmacy students have the opportunity to improve their communication skills with people of all ages at outreach events.

Gaining Cultural Competence

Providing blood pressure readings, diabetes risk assessments, and vision screenings at health fairs in parks and hospitals allows pharmacy students to practice their communication skills with, and develop understanding of, people of different cultures and backgrounds.

Finding a Faculty Mentor

Besides teaching pharmacy students in classrooms, faculty members often practice and precept at sites in the community. By learning from and observing faculty members who may attend community service events, pharmacy students can develop mentor–mentee relationships. That connection can turn into an opportunity to conduct research or receive a glowing letter of recommendation.

Further Developing Drug Information and Patient Counseling Skills

Pharmacy students can gain experience with finding drug interactions, assessing patient adherence, and counseling patients on medications and other health-related issues through volunteering.

Developing New Friendships

Volunteering with other pharmacy students makes the experience more enjoyable, as pharmacy students can take a break from studying to come together to speak to patients, make a difference in people's lives, and perhaps go out together for a treat afterwards. Pharmacy students who gather with the same passion for community service may develop strong bonds and friendships that can help them grow academically and emotionally—whether it be through partnering on research projects, sharing study notes, or supporting each other through the rigors of pharmacy school.

Community Service Opportunities Outside of the Pharmacy

Pharmacy students should be aware of the plethora of volunteer opportunities available in their own communities. Volunteering within the medical sector is an outstanding opportunity and accomplishment, but it is not feasible for everyone. Volunteering at non–pharmacy-related community service events is just as effective in fostering personal growth. Volunteering at community service events will help pharmacy students become better pharmacists and better people.

11.3 Online Community Services Tools and Ideas

11.3.1 YouTube

YouTube can be used effectively in online community services. YouTube can be used for streaming to the audience and having a one-way video interaction. YouTube can be a good resource for educational and awareness videos. Pharmacy educators and students can use YouTube for recording lectures and sharing them with the public where the public and patients can easily access material at any time.

11.3.2 Social Media

Social media can be used effectively in online community services in many ways as follows: communication tools between the pharmacy educators/students and the public; sharing educational and awareness materials such as brochures, videos, and posts.

11.3.3 WhatsApp

WhatsApp can be used effectively in online community services. Pharmacy educators and students can share educational and awareness materials such as brochures, videos, and posts with their relatives, friends, and the public.

11.3.4 Awareness Campaigns

Pharmacy schools/departments play a very important role in international awareness campaign days such as diabetes awareness day, breast cancer awareness day, and others. Furthermore, many pharmacy schools/departments around the world have designed their own awareness campaigns based on society needs such as medication safety awareness day.

11.3.5 Drug (Medications) and Poison Information Services

Pharmacy schools/departments play a very important role in providing drug (medications) and poison information services to patients, the public, and health care professionals through websites, mobile applications, social media, and others.

11.4 Conclusion

Volunteering in community services allows pharmacy students to give back to their communities, develop skills needed for practice in an ambulatory care or community setting, gain insight into the lives of people who are underserved, and learn about people of different cultures and backgrounds.

References

Accreditation Council for Pharmacy Education (ACPE), 2015. *Accreditation standards and key elements for the professional program in pharmacy leading to the doctor of pharmacy degree* (Standards 2016). Available at: https://www.acpe-accredit.org/pdf/Standards2016FINAL.pdf

Farzadeh, S., 2019. The role of community service in building better pharmacists. *American Journal of Health-System Pharmacy*, 76, pp. 644–645.

12

Access and Equitable Access

12.1 Background

Access and equitable access to higher education in general are major challenges facing students in many middle- and low-income countries and perhaps many students in high-income countries. Online education became a factor in higher education after the lockdown during the COVID-19 pandemic, as the majority of universities were closed during the lockdown and education moved to a remote/online style. However, it was very difficult to adapt and implement the distance education model in pharmacy education undergraduate programs as the nature of the study at pharmacy schools' programs requires many laboratories and training at schools, primary health care facilities, secondary health facilities, and tertiary health care facilities. However, after the successful implementation of the new technologies in the early 1990s by the Southeastern University of the Health Sciences, which developed a North Miami site to service its non-traditional, post baccalaureate, Doctor of Pharmacy (PharmD) program in 1991 and forwards worldwide, online pharmacy education has changed towards adapting and implementing many programs, and therefore offer online degrees for pharmacy students, especially for postgraduate studies. Since early 2020, the majority of pharmacy schools/departments have offered all degrees at both undergraduate and postgraduate levels completely online or as a hybrid model (theory courses delivered online, with training in hospitals or pharmacies). Access as well as equitable access to pharmacy education and online pharmacy education is different from one country to another and is affected by many factors and issues as follows:

Cost of Pharmacy Education

The cost of pharmacy education is different from one country to another and even from one school to another in the same country. There are different education systems around the world. While pharmacy education is free in many public schools in many countries, it is very costly in other countries. While many countries offer scholarships to support education in general and

DOI: 10.1201/9781003230458-13

pharmacy education in particular, students suffer in many countries around the world to learn pharmacy. Private schools are expensive in general in developed countries and are more expensive in the developing countries. The cost of pharmacy education plays a very important role in the access and equitable access to pharmacy education.

Number of Pharmacy Schools/Departments

The number of pharmacy schools/departments is different from one country to another. While there is an adequate number of pharmacy schools/departments in many countries, there is an inadequate number of pharmacy schools/departments in many countries around the world, especially in the developing countries. The number of pharmacy schools/departments plays a very important role in the access and equitable access to pharmacy education. However, with online education, this issue could be solved by a higher, more adequate number of pharmacy educators and staff.

Location of Pharmacy Schools/Departments

The location of pharmacy schools/departments plays a very important role in the access and equitable access to pharmacy education. In many countries around the world, the location of pharmacy schools is far away from the majority of students' cities and villages, which makes admission to pharmacy schools not affordable to them. However, with online education, this issue could be solved.

Internet

The internet plays a very important and vital role in online pharmacy education. The internet facilitates the teaching and learning process which makes distance/online pharmacy education more effective and easier than at any time in history. Pharmacy educators and students need the internet for: making communication easy; delivering classes; uploading/downloading lecture notes and other educational materials/resources; exams; assignments; presentations; searching for pharmacy-related information; training; patient care services; public health promotion, awareness, and services. Without access to the internet, online teaching and learning will stop. Access and equitable access to the internet varies among countries, and even among educators and students in the same country. Internet cost is expensive in many countries around the world and many pharmacy educators and students are not able to pay the monthly cost of internet in many middle- and low-income countries, especially in the developing countries. Many areas, cities, and rural areas don't have access to the internet in many middle- and low-income countries in the developing countries. Quality and speed of internet play a very important role in access and equitable access to online pharmacy education. Policy makers should make efforts and develop solutions to this issue to make access to online pharmacy education possible to all educators and students.

Computers and Laptops

Using computers and laptops is important and essential for online teaching and learning. It provides flexible and effective access to online teaching and learning for pharmacy educators and students. Access and equitable access to computers and laptops varies among countries, and even among educators and students in the same country. Computers and laptops are expensive in many countries around the world and many pharmacy educators and students are not able to pay for them, especially in many developing countries. Policy makers should make efforts to develop solutions to this issue to make access to online pharmacy education possible to all educators and students. Computer and laptop companies should also contribute to solving this issue in developing countries. Private sectors should also contribute to solve this issue.

Smartphones, Tablets, and Net Books

Using smartphones, tablets, and net books is very important even essential for online teaching and learning. It provides flexible and effective access to online teaching and learning for pharmacy educators and students. It helps students to download many education resources and others. Furthermore, students can access it at any time throughout the day inside or outside of home. It helps educators also to prepare, revise, and access lecture notes, literature, educational resources, and others at any time and from any place. Access and equitable access to smartphones, tablets, and net books varies among countries, and even among educators and students in the same country. Smartphones, tablets, and net books are expensive in many countries around the world and many pharmacy educators and students are not able to pay for them, especially in many developing countries. Policy makers should make efforts to develop solutions to this issue to make access to online pharmacy education possible for all educators and students. Smartphones, tablets, and net books companies should also contribute to solving this issue in developing countries. Private sectors should also contribute to solving this issue.

Learning Management Systems (LMS)

Learning Management Systems (LMS) are very important in online pharmacy education as well as in higher education. They contain an effective web-based learning system of sharing study materials, making announcements, conducting evaluation and assessments, generating results, and communicating interactively in synchronous and asynchronous ways, among various other academic activities (Kant et al., 2021; Bervell and Umar, 2017). Pharmacy schools in low-income countries can use the free Learning Management Systems (LMS) for the teaching and learning process.

Webinar and Video Conferencing Platforms

Webinar and video conferencing platforms are very important in online pharmacy education. Microsoft Teams, Cisco WebEx Teams, Google Meet, and Zoom are the most common webinar and video conferencing platforms in online pharmacy education (www.microsoft.com/en-us/microsoft-teams/group-chat-software; www.webex.com/; https://apps.google.com/meet/; www.zoom.us/).

Microsoft Teams, Cisco WebEx Teams, Google Meet, and Zoom offer multiple versions of their software based on usage requirements. This includes free versions that are great for light uses, short conference calls, and light file sharing. However, universities may pay the cost of all platforms for pharmacy educators and students to facilitate teaching and learning. All platforms are used successfully in online pharmacy education worldwide. Training, workshops, and writing manuals are very important for both pharmacy educators and students in order to use these software tools successfully. The majority of pharmacy schools worldwide have no problem with this issue; however, many pharmacy schools in low-income countries cannot afford the cost of unlimited versions. Microsoft Teams, Cisco WebEx Teams, Google Meet, and Zoom can contribute and provide free versions to pharmacy schools in very low- and low-income countries as a contribution to society and this will also help to achieve access and equitable access to online pharmacy education.

Online and Digital Library

An online and digital library is a collection of documents such as journal articles, books, and other educational resources organized in an electronic form and available on the internet for students, educators, and other staff. Furthermore, access to databases helps students, educators, and staff to access the latest volumes/issues of scientific journals. The majority of pharmacy schools worldwide have no problem with this issue; however, many pharmacy schools in low-income countries cannot afford the cost of online digital libraries and they don't have the resources for launching one. International publishers can help the pharmacy schools in very low- and low-income countries to solve this issue and give them free access.

Wearable Technologies

Wearable technologies have a tremendous potential to improve education, empowering students as well as instructors in their teaching and learning experiences (Borthwick et al., 2015; Subrahmanyam and Swathi, 2020). Wearable technologies can be used in online pharmacy education with the following potential reasons: improved student engagement; convenient to wear as they are hands-free gadgets; effortless communication with enhanced features; facility to record videos; teaching through augmented reality; teaching through virtual reality and wearables–learning apps (Subrahmanyam and Swathi, 2020). Access to wearable technologies varies from one country to another, and even among pharmacy educators and students in the same country or in the same pharmacy school/department. There are many factors affecting access and equitable access to wearable technologies.

Practicals/Laboratories and Tutorials

Technology has played an important role in education which has led to replacing or decreasing the need for the traditional laboratories (labs) such as adapting simulation for pharmacology, anatomy, and other subjects, which leads to improving the learning outcomes and, moreover, saving the pharmacy schools' budgets. However, not all pharmacy schools/departments, especially in developing countries, have adapted simulation to replace certain laboratories.

Learning Environments at Home

Learning environment is very important in online learning, and not all pharmacy educators or students have suitable environments at home, which affects equitable access to online pharmacy education.

Digital and Technology Literacy

Digital media, technologies, and online tools are very important for online pharmacy education; therefore, it is essential that pharmacy educators as well as students develop their digital and technology literacy, which is different among pharmacy educators and students. Pharmacy schools/departments should make efforts to solve this issue in order to train all pharmacy educators and students about the necessary technologies for online pharmacy education.

Workforce Issues

The number of pharmacy educators, teaching assistants, and staff are different from one pharmacy school to another, and from public schools to private schools. While there are adequate pharmacy educators, teaching assistants, and staff in the majority of public schools, there is a lack of pharmacy educators, teaching assistants, and staff in the private schools, especially in many developing countries. An adequate number of trained pharmacy educators, teaching assistants, and staff is very important in pharmacy education as well as online pharmacy education for the theory courses, practicals, tutorials, training, community services, and research.

Support

Support with training, workshops, and technical support for pharmacy educators as well as students is important for the success of online pharmacy education. There are differences in support between countries, and from one school to another. Pharmacy schools/departments should make efforts to solve this issue in order to provide the necessary and required support and training to all pharmacy educators and students for the success of online pharmacy education.

12.2 Conclusion

Access and equitable access to pharmacy education and online education are major challenges facing students in many middle- and low-income countries and perhaps many students in high- income countries. There are many factors affecting access and equitable access to pharmacy education and online education such as the cost of pharmacy education, the lack of technologies, and others.

References

Bervell, B. and Umar, I.N., 2017. A decade of LMS acceptance and adoption research in Sub-Sahara African higher education: A systematic review of models, methodologies, milestones and main challenges. *EURASIA Journal of Mathematics, Science and Technology Education*, 13(11), pp. 7269–7286.

Borthwick, A.C., Anderson, C.L., Finsness, E.S. and Foulger, T.S., 2015. Special article personal wearable technologies in education: Value or villain? *Journal of Digital Learning in Teacher Education*, 31(3), pp. 85–92.

Kant, N., Prasad, K.D. and Anjali, K., 2021. Selecting an appropriate learning management system in open and distance learning: a strategic approach. *Asian Association of Open Universities Journal*, 16(1), pp. 79–97.

Subrahmanyam, V.V. and Swathi, K., 2020. *Wearable technology and its role in education*. International Conference—2020 on Distance Education and Educational Technology (ICE-CODL 2020), CDOL, JMI, New Delhi 10–11Dec'2020

13

Quality and Accreditation

13.1 Importance of Accreditation in Pharmacy Education and Online Pharmacy Education Quality

Accreditation is very important to ensure that pharmacy schools/departments are able to graduate students with the essential competencies to be able to provide effective pharmacist care and patient care services face to face or online. Local and international accreditations agencies go to great efforts to supervise pharmacy education programs and monitor them, license their programs, and relicense their programs every 3 to 5 years to be sure future pharmacists receive the appropriate education and achieve the necessary competencies in terms of (ACPE, 2015; GPC, 2021; NAPRA/ANORP, 2014; Rouse and Meštrovic, 2014; Sacre et al., 2020; APC, 2020):

Knowledge (learner, educator, and health and wellness promoter and counselor)

Cognitive (thinker, analyzer, problem solver, and decision maker)

Communication, education, and collaboration (communicator, educator, and collaborator)

Life-long learning and personal/professional development (learner and innovator)

Leadership and management (leader)

Pharmaceutical marketing, pharmacist care marketing (promoter)

Pharmaceutical industry (manufacturers)

Pharmacist and patient care (care provider and innovator)

Medication safety (care provider and health protector)

Prescribing (prescriber)

Ethical, legal and professional responsibilities

Health promotion and community services

DOI: 10.1201/9781003230458-14

Research (researcher)

Technology (IT user)

Please refer to Chapter 4: Competencies and Learning Outcomes for more details.

13.2 What Is Accreditation?

Vlăsceanu et al. (2004) defines accreditation as: the process by which a (non-)governmental or private body evaluates the quality of a higher education institution as a whole or of a specific educational program in order to formally recognize it as having met certain predetermined minimal criteria or standards. The result of this process is usually the awarding of a status (a yes/no decision), of recognition, and sometimes of a license to operate within a time-limited validity. The process can imply initial and periodic self-study and evaluation by external peers. The accreditation process generally involves three specific steps: (i) a self-evaluation process conducted by the faculty, the administrators, and the staff of the institution or academic program, resulting in a report that takes as its reference the set of standards and criteria of the accrediting body; (ii) a study visit, conducted by a team of peers, selected by the accrediting organization, which reviews the evidence, visits the premises, and interviews the academic and administrative staff, resulting in an assessment report, including a recommendation to the commission of the accrediting body; (iii) an examination by the commission of the evidence and recommendation on the basis of the given set of criteria concerning quality and resulting in a final judgment and the communication of the formal decision to the institution and other constituencies, if appropriate (Vlăsceanu et al., 2004).

13.3 History of Accreditation

The history of accreditation in general goes back to 1784 when the New York Board Regents became the first accrediting agency in the US (Brady, 1988).

13.3.1 History of Medical Education Accreditation

The history of medical education accreditation goes back to the 1840s (American Medical Association, 1983).

13.3.2 History of Pharmacy Education Accreditation

The history of pharmacy education accreditation goes back to 1932 when the American Council on Pharmaceutical Education (ACPE) was launched; the agency's name was changed to the Accreditation Council for Pharmacy Education (ACPE) in 2003. The ACPE initially established standards for the baccalaureate degree in pharmacy and then added the Doctor of Pharmacy standards as an alternative. In 2000, the ACPE announced the conversion to the Doctor of Pharmacy (PharmD) as the sole entry-level degree for the profession of pharmacy. In 1975, the ACPE developed standards for the approval (now, accreditation) of providers of continuing pharmacy education and in 1999, developed additional standards for CE providers who were conducting certificate programs in pharmacy (AIHP, 2018; ACPE, 2021).

The ACPE's International Services Program (ISP) was established in February 2011. The program was created to strengthen the ACPE's ability to assist international stakeholders who seek guidance related to quality assurance and advancement of pharmacy education.

13.4 Local Pharmacy Education Accreditation around the World

Pharmacy programs' accreditation systems, bodies, regulations, and standards are different around the world. While many countries established accreditation bodies such as the Accreditation Council for Pharmacy Education (ACPE) in the United States, there are many countries without accreditation systems and the approval of programs are from the Ministries of Higher Education/ Education; however, this does not mean that the quality of education is poor. The most important issue that the accreditation bodies or ministries should ensure is to monitor the quality of pharmacy education in their countries. Pharmacy regulation authorities play an important role in the accreditation of pharmacy programs in many countries around the world.

13.5 International Accreditation

Pharmacy schools/departments aim to graduate students who are competent and able to work and provide pharmacist care services and patient care services inside their countries as well as outside their countries. Many pharmacy schools around the world have obtained international accreditation and certification for their programs such as the Accreditation Council for Pharmacy Education (ACPE), US, and The Canadian Council for Accreditation of Pharmacy Programs.

13.6 Accreditation Standards

Accreditation standards could be a little different from one accreditation body to another. Examples of accreditation standards are as follows:

Accreditation Council for Pharmacy Education (ACPE), accreditation standards for PharmD (ACPE, 2015):

Standard 1: Foundational Knowledge

Standard 2: Essentials for Practice and Care

Standard 3: Approach to Practice and Care

Standard 4: Personal and Professional Development

Standard 5: Eligibility and Reporting Requirements

Standard 6: College or School Vision, Mission, and Goals

Standard 7: Strategic Plan

Standard 8: Organization and Governance

Standard 9: Organizational Culture

Standard 10: Curriculum Design, Delivery, and Oversight

Standard 11: Interprofessional Education (IPE)

Standard 12: Pre-Advanced Pharmacy Practice Experience (Pre-APPE) Curriculum

Standard 13: Advanced Pharmacy Practice Experience (APPE) Curriculum

Standard 14: Student Services

Standard 15: Academic Environment

Standard 16: Admissions

Standard 17: Progression

Standard 18: Faculty and Staff—Quantitative Factors

Standard 19: Faculty and Staff—Qualitative Factors

Standard 20: Preceptors

Standard 21: Physical Facilities and Educational Resources

Standard 22: Practice Facilities

Standard 23: Financial Resources

Standard 24: Assessment Elements for Section I: Educational Outcomes

Standard 25: Assessment Elements for Section II: Structure and Process

13.7 Online Pharmacy Education Accreditation

It is very difficult to adapt and implement distance education and online education models in pharmacy education undergraduate programs such as Bachelor of Pharmacy or Doctor of Pharmacy as the nature of the study at pharmacy schools' programs requires many laboratories, and training at schools, primary health care facilities, secondary health facilities, and tertiary health care facilities.

However, after the successful implementation of the new technologies in the early 1990s by the Southeastern University of the Health Sciences, which developed a North Miami site to service its non-traditional, post baccalaureate, Doctor of Pharmacy (PharmD) program in 1991 and forwards worldwide, online pharmacy education has changed towards adapting and implementing many programs, and therefore offer online degrees for pharmacy students, especially for postgraduate studies. Since early 2020, the majority of pharmacy schools/departments have offered all degrees at both undergraduate and postgraduate levels completely online or as a hybrid model (theory courses delivered online, with training in hospitals or pharmacies). However, nowadays after the successful implementation of the new technologies in the early 1990s in pharmacy education, pharmacy schools/departments can offer many postgraduate programs completely online, especially postgraduate diploma and master programs. Each country accreditation body should develop clear guidelines and standards about the accreditation of all online pharmacy programs and certificates, especially at the postgraduate level.

13.8 Achievements

Nowadays, the majority of pharmacy schools/departments in developed countries as well as in many developing countries are accredited locally and many internationally.

13.9 Conclusion

Accreditation is very important to ensure that pharmacy schools/departments are able to graduate students with the essential competencies to be able to provide effective pharmacist care and patient care services face to face or online. Local and international accreditations agencies go to great efforts to supervise the pharmacy education programs and monitor them, license their programs, and relicense their programs every 3 to 5 years to be sure future pharmacists receive the appropriate education and achieve the necessary competencies.

References

Accreditation Council for Pharmacy Education (ACPE), 2015. *Accreditation standards and key elements for the professional program in pharmacy leading to the doctor of pharmacy degree* (Standards 2016). Available at: www.acpe-accredit.org/pdf/Standards2016FINAL.pdf

Accreditation Council for Pharmacy Education (ACPE), 2021. *About: Accreditation council for pharmacy education.* Available at: acpe-accredit.org.

AIHP., 2018. *Guidelines on teaching history in pharmacy education.* Available at: https://aihp.org/wp-content/uploads/2018/12/AACP-D.pdf

American Medical Association, 1983. History of accreditation of medical education programs. *JAMA*, 250(12), pp. 1502–1508.

Australian Pharmacy Council (APC), 2020. www.pharmacycouncil.org.au/

Brady, J.E., 1988. Accreditation: A historical overview. *Hospitality & Tourism Educator*, 1(1), pp. 18–24.

General Pharmaceutical Council (GPC), 2021. *Standards for pharmacy education.* Available at: www.pharmacyregulation.org/sites/default/files/document/standards-for-the-initial-education-and-training-of-pharmacists-january-2021_0.pdf

NAPRA (National Association of Pharmacy Regulatory Authorities)/ANORP (Association nationale des organismes de réglementation de la pharmacie)., 2014. *Professional competencies for canadian at entry to practice pharmacists.* Available at: https://napra.ca/sites/default/files/2017-08/Comp_for_Cdn_PHARMACISTS_at_EntrytoPractice_March2014_b.pdf

Rouse, M. and Meštrovic, A., 2014. *Quality assurance of pharmacy education: The FIP Global Framework.* International Pharmaceutical Federation (FIP).

Sacre, H., Hallit, S., Hajj, A., Zeenny, R.M., Akel, M., Raad, E. and Salameh, P., 2020. Developing core competencies for pharmacy graduates: The Lebanese experience. *Journal of Pharmacy Practice*, p. 0897190020966195.

Vlăsceanu, L., Grünberg, L. and Pârlea, D., 2004. *Quality assurance and accreditation: A glossary of basic terms and definitions* (p. 25). Unesco-Cepes.

14

Advantages and Disadvantages

14.1 Advantages of Online Pharmacy Education

14.1.1 Supporting the Continuity of Teaching and Learning

Online education provides innovative and resilient solutions in times of crisis to pharmacy education worldwide. The majority of pharmacy schools/departments worldwide have adapted and implemented online education successfully with the help of technologies and new technologies, which saved the teaching and learning process during the COVID-19 lockdown. Otherwise, pharmacy education would have stopped for 1 year or more because of the pandemic. However, many low-income countries couldn't adapt or implement online education due to many reasons.

14.1.2 Cost Saving

Pharmacy schools can save money with online teaching as can students with online learning. Saving indirect costs such as the cost of transportation, accommodation, actual lab sessions, and so on can help pharmacy schools' budgets as well as students.

14.1.3 Convenience

Online pharmacy education from home can be convenient for many pharmacy educators and students.

14.1.4 Improving Technology Knowledge

Online pharmacy education is a good opportunity for pharmacy educators and students to improve their knowledge of technologies, including new technologies.

DOI: 10.1201/9781003230458-15

14.1.5 Improving Technology Skills

Online pharmacy education is a good opportunity for pharmacy educators and students to improve their skills in technologies, including new technologies.

14.1.6 Self-Learning

Online pharmacy education is good opportunity for pharmacy students to improve their self-learning related skills to be good pharmacists with life-long learning skills.

14.1.7 Accessible and Affordable

Online pharmacy education is a good opportunity for those who are currently working or who can't travel to obtain their postgraduate studies. A good example of this is online master's degrees, PharmD, for working pharmacists and others.

14.1.8 Schedule

Online pharmacy education can help pharmacy schools to arrange schedules with more flexibility, especially for the labs and tutorials.

14.1.9 Safety

Online pharmacy education is safer than face-to-face education, with pharmacy educators and students teaching and learning at their homes, which helps in reducing the contacts and managing and control of the spread of COVID-19 as an example. There are no labs, which means there is no potential risk due to chemicals and other factors.

14.1.10 Infrastructure

Online pharmacy education requires less infrastructure than traditional education.

14.1.11 Teaching Strategies

Online pharmacy education is a good opportunity to apply and practice more active teaching strategies.

14.1.12 Assessment and Evaluation Methods

Online pharmacy education is a good opportunity to apply and practice online assessment and evaluation methods, explore the barriers, and make solutions.

14.1.13 Community Services

Online pharmacy education is a good opportunity to apply and practice online community services for the public and patients.

14.1.14 Environment

Online pharmacy education has benefits for the environment.

14.1.15 Workforce Issues

Pharmacy schools can hire many international faculty members as part-time, visiting, and adjunct professors and lecturers more easily.

14.1.16 Communication and Collaboration

Online pharmacy education is a good opportunity to practice online/virtual communication and collaboration, which improves the skills of communication and collaboration.

14.1.17 Time Management

Online pharmacy education has benefits for pharmacy educators and students related to time management.

14.2 Disadvantages and Problems of Online Pharmacy Education

Online pharmacy education has many disadvantages and problems such as the following:

Students' Engagement

Many students faced this problem due to many causes such as home size and other factors.

Lack of Infrastructure and Resources

Many pharmacy schools around the world, especially in low-income and middle-income countries, faced this problem such as the following:

Internet

The internet plays a very important and vital role in online pharmacy education. The internet facilitates the teaching and learning process, which makes distance/online pharmacy education more effective and easier than at any time in history. Pharmacy educators and students need the internet for: making communication easy; delivering the classes; uploading/downloading the lecture notes and other educational materials/resources; exams; assignments; presentations; searching for pharmacy-related information; training; patient care services; public health promotion, awareness, and services. Without access to the internet, online teaching and learning will stop. Many pharmacy educators and students, especially in low-income and middle-income countries, face this problem.

Computers and Laptops

Using computers and laptops is essential for online teaching and learning. It provides flexible and effective access to online teaching and learning for pharmacy educators and students. Many pharmacy educators and students, especially in low-income and middle-income countries, face the problem of lack of access.

Smartphones, Tablets, and Net Books

Using smartphones, tablets, and net books is very important and essential for online teaching and learning. It provides flexible and effective access to online teaching and learning for pharmacy educators and students. It helps students to download many education resources and other materials. Furthermore, they can access it at any time throughout the day inside or outside home. It helps educators also to prepare, revise, and access the lecture notes, literature, educational resources and other materials at any time and any place. Many pharmacy educators and students, especially in low-income and middle-income countries face the problem of lack of access.

Online and Digital Library

An online and digital library is a collection of documents such as journal articles, books, and other educational resources organized in an electronic form and available on the internet for students, educators, and other staff. Furthermore, it provides access to the databases which help students, educators, and staff to access the latest volumes/issues of scientific journals. Many pharmacy educators and students, especially in low-income and middle-income countries face the problem of lack of access.

Lack of/Poor Communication Skills

Many pharmacy educators as well as students have this problem.

Lack of/Poor Technology Use Skills

Many pharmacy educators as well as students have this problem.

Assessment and Evaluation

There are many problems related to assessment and evaluation such as plagiarism, cheating, and others.

Workload

Many pharmacy educators' workload has been increased due to many reasons such as more activities, more meetings, and others.

Lack of/Insufficient Training

Many pharmacy educators as well as students face this problem.

Lack of/Insufficient Technical Support

Many pharmacy educators as well as students face this problem.

Quality of Online Education

This is a problem in many countries around the world, especially in low-income and middle-income countries.

Lack of Self-Motivation

Many pharmacy educators as well as students face this problem.

Distraction

Many pharmacy educators as well as students face this problem.

Attitude

Many pharmacy educators as well as students have a negative attitude towards online learning.

14.3 Conclusion

Online education provides innovative and resilient solutions to pharmacy education worldwide in times of crisis. The majority of pharmacy schools/

departments worldwide adapted and implemented online education successfully with the help of technologies and new technologies, which saved the teaching and learning process during the COVID-19 lockdown. This chapter describes the advantages and disadvantages of online pharmacy education.

Section 2

Online Pharmacy Practice

15

History and Importance

15.1 History of Pharmacy Practice

It is reported that pharmacy practice history goes back to ancient times, when old men used natural resources such as herbs for their health (Bender and Thom, 1952). Historical reports exist of the following examples of development in pharmacy education around the world (Bender and Thom, 1952; Al-Ghazal and Tekko, 2003):

Pharmacy in Ancient Babylonia

Babylon, jewel of ancient Mesopotamia, often called the cradle of civilization, provides the earliest known record of practice of the art of the apothecary. Practitioners of healing of this era (about 2600 BC) were priest, pharmacist, and physician, all in one. Medical texts on clay tablets record first the symptoms of illness, the prescription and directions for compounding, then an invocation to the gods. Ancient Babylonian methods find their counterpart in today's modern pharmaceutical, medical, and spiritual care of the sick.

Pharmacy in Ancient China

Chinese pharmacy, according to legend, stems from Shen Nung (about 2000 BC), an emperor who sought out and investigated the medicinal value of several hundred herbs.

Pharmacy in Ancient Egypt (Days of the Papyrus Ebers)

Though Egyptian medicine dates from about 2900 BC, its best known and most important pharmaceutical record is the "Papyrus Ebers" (1500 BC), a collection of 800 prescriptions, mentioning 700 drugs. Pharmacy in ancient Egypt was conducted by two or more echelons: gatherers and preparers of drugs, and "chiefs of fabrication," or head pharmacists. They are thought to have worked in the "House of Life." In a setting such as this, the "Papyrus

Ebers" might have been dictated to a scribe by a head pharmacist as he directed compounding activities in the drug room.

Theophrastus—Father of Botany

Theophrastus (about 300 BC), among the greatest early Greek philosophers and natural scientists, is called the "father of botany."

The First Apothecary Shops

The Arabs separated the arts of apothecary and physician, establishing in Bagdad late in the eighth century the first privately owned drug stores.

Separation of Pharmacy and Medicine

In European countries exposed to Arabian influence, public pharmacies began to appear in the 17th century. However, it was not until about 1240 AD that, in Sicily and southern Italy, pharmacy was separated from medicine.

Pharmacopoeia

Sabur Ibn Sahl (d. 869), was, however, the first physician to initiate a pharmacopoeia, describing a large variety of drugs and remedies for ailments. Al-Biruni (973–1050) wrote one of the most valuable Islamic works on pharmacology entitled Kitab al-Saydalah (The Book of Drugs), where he gave detailed knowledge of the properties of drugs and outlined the role of pharmacy and the functions and duties of the pharmacist. Ibn Sina (Avicenna), too, described no less than 700 preparations, their properties, modes of action, and their indications.

Pharmacy Practice Today

Pharmacy practices in developed and many developing countries have witnessed many changes during the last decades towards improving pharmacist care/patient care services. The revolutionary changes in the pharmacy practice a few decades ago, when the pharmacy profession witnessed great practice changes and moved away from its original focus on medicine supply and dispensing towards a focus on patient care, especially after the introduction of clinical pharmacy concepts in the late 1960s, followed by the philosophy of pharmaceutical care in the early 1990s. However, pharmacists nowadays play/should play an important role in patient care. They contribute effectively to patients' health; diseases/conditions management as well as prevention; treating outcomes; treating cost; patients' quality of life; and satisfaction towards the health care system and care.

15.2. History of Distance/Remote and Online Pharmacy Practice

15.2.1 Mail-Order Pharmacy

The history of distance and remote pharmacy services goes back to the early 1940s in the United States with using mail order medicines as follows (Nicole, 2020; Horgan et al., 1990):

1946

In 1946, the United States saw the first pharmacy dedicated to mailing prescription drugs to a patient's home. The Veterans Administration (VA) was the first to offer services to eligible veterans. Today, the VA still accounts for nearly one-third of the mail-order prescriptions that are dispensed in the United States.

1959

The American Association of Retired Persons and the National Retired Teachers Association formed a mail pharmacy that was not-for-profit and served only their membership. By 1963, for-profit entities began to market mail-order pharmacy services to corporate, government, and union employers. There was some opposition to this type of pharmacy throughout the 1960s.

1980s

The 1980s saw the most rapid growth of the mail-order pharmacy industry with revenues soaring from $100 million to $1.5 billion. The trend would continue to grow throughout the 1990s mirroring the growth of the internet. Mail-order pharmacy became more cost effective and convenient.

2000s

Mail-order pharmacy services have increased in many countries around the world.

15.2.2 Drug Information and Counseling Services

The history of drug information centers goes back to the early 1960s (Parker, 1965; Gabay, 2017). Pharmacists have used telephones for patient education and counseling since the 1960s (Lester, 1977).

15.2.3 Telepharmacy

Telepharmacy is the delivery of pharmaceutical care via telecommunications to patients in locations where they may not have direct contact with a pharmacist. The history of telepharmacy goes back to 1942. Telepharmacy has been in use by Australia's Royal Flying Doctor Service since 1942 (Margolis and Ypinazar, 2008). Since the 2000s many countries around the world have launched telepharmacy services.

15.2.4 Online Pharmacies

Literature reports that online pharmacies history goes back to 1999 and 2000s after the evolution of the internet. Internet pharmacy is also known as online pharmacy, cyberpharmacy, e-pharmacy, and virtual pharmacy/drugstores. The introduction of Soma.com marked the arrival of a major pharmacy presence on the internet in January 1999 (Gallagher and Colaizzi, 2000). Many online pharmacy sites followed soon after, with drugstore.com and PlanetRx. com being the most notable. These three sites were widely regarded as among the most credible and reputable in the industry. Most internet pharmacy sites were initially stand-alone, full-service online pharmacies. The industry has shifted quickly, with traditional pharmacy chains aggressively acquiring all or part of the internet start-ups or establishing their own divisions for online prescription and nonprescription sales (Crawford, 2003).

15.3 Importance of Distance/Remote and Online Pharmacist Care

Distance/remote and online pharmacist care has many benefits to patients and the public, especially those in rural areas or those who cannot access or reach the pharmacies and pharmacists. There are more benefits such as lower cost, convenience, and greater anonymity for consumers.

15.4 Conclusion

This chapter has discussed the history and importance of distance and online pharmacy practice. This chapter includes the background of pharmacy practice, development, and importance. It also includes the history of distance and online practice.

References

Al-Ghazal, S.K. and Tekko, I.A., 2003. The valuable contributions of Al-Razi (Rhazes) in the history of pharmacy during the Middle Ages. *Journal of the International Society for the History of Islamic Medicine*, 2(9), pp. 9–11.

Bender, G.A. and Thom, P.B.R.A., 1952. *A history of pharmacy in pictures*. Davis & Company.

Crawford, S.Y., 2003. Internet pharmacy: Issues of access, quality, costs, and regulation. *Journal of Medical Systems*, 27(1), pp. 57–65.

Gallagher, J.C. and Colaizzi, J.L., 2000. Issues in Internet pharmacy practice. *Annals of Pharmacotherapy*, 34(12), pp. 1483–1485.

Gabay, M.P., 2017. The evolution of drug information centers and specialists. *Hospital Pharmacy*, 52(7), pp. 452–453.

Horgan, C., Goody, B., Knapp, D. and Fitterman, L., 1990. The role of mail service pharmacies. *Health Affairs*, 9(3), pp. 66–74.

Lester, D., 1977. The use of the telephone in counseling and crisis intervention. *The Social Impact of the Telephone*, pp. 454–472.

Margolis, S. and Ypinazar, V., 2008. *Tele-pharmacy in remote medical practice: The royal flying doctor service medical chest program*. Available at: https://search.informit.org/doi/abs/10.3316/informit.467724863149788

Nicole, Kruczek., 2020. *What is the role of the mail-order pharmacy?* Available at: www.pharmacytimes.com/view/what-is-the-role-of-the-mail-order-pharmacy

Parker, P.F., 1965. The University of Kentucky drug information center. *American Journal of Health-System Pharmacy*, 22(1), pp. 42–47.

16

Online Pharmacies

16.1 Online Pharmacies Definition

An online pharmacy, internet pharmacy, or mail-order pharmacy is a pharmacy that operates over the internet and sends orders to customers through mail, shipping companies, or an online pharmacy web portal (Online pharmacy, Wikipedia, 2021). The General Pharmaceutical Council defines an internet pharmacy as "a registered pharmacy which offers to sell or supply medicines (or other pharmaceutical products) and/or provides other professional services over the internet, or makes arrangements for the supply of such products or provision of such services over the internet" (Rickert and Anderson, 2000).

16.2 History of Pharmacy

It is reported that the pharmacy practice history goes back to ancient times, when old men used natural resources such as herbs for their health (Bender and Thom, 1952). Historical records show the following examples of development in pharmacy around the world (Bender and Thom, 1952; Al-Ghazal and Tekko, 2003).

16.3 History of Mail Order Pharmacy

The history of distance and remote pharmacy services goes back to the early 1940s in the United States with using mail order medicines as follows (Nicole, 2020; Horgan et al., 1990): In 1946, the United States saw the first pharmacy dedicated to mailing prescription drugs to a patient's

DOI: 10.1201/9781003230458-18

home. The Veterans Administration (VA) was the first to offer services to eligible veterans. Today, the VA still accounts for nearly one-third of the mail-order prescriptions that are dispensed in the United States. In 1959, The American Association of Retired Persons and the National Retired Teachers Association formed a mail-order pharmacy that was not-for-profit and served only their membership. By 1963, for-profit entities began to market mail-order pharmacy services to corporate, government, and union employers. There was some opposition to this type of pharmacy throughout the 1960s.

The 1980s saw the most rapid growth of the mail-order pharmacy industry with revenues soaring from $100 million to $1.5 billion. The trend would continue to grow throughout the 1990s mirroring the growth of the internet. Mail-order pharmacy became more cost effective and convenient. In the 2000s, mail-order pharmacy services have increased in many countries around the world.

16.4 Characteristics of Online Pharmacies

Literature reports that online pharmacies should perform the following desirable activities (Kumaran et al., 2020; Orizio et al., 2011): (1) Require a prescription from a licensed prescriber to purchase prescription-only drugs. Online pharmacies provide easy access to traditionally controlled substances, such as pharmaceutical products, which in regulated systems need an original medical prescription before they can be purchased. (2) Evaluate consumer health status via online questionnaires prior to drug purchases. Some online pharmacies require a questionnaire to be completed before a purchase can be made, which helps them store consumer information. (3) Provide contact details for the online pharmacy. Although this is crucial information that websites should provide, some online pharmacies currently refuse to provide their contact details. (4) Declare the geographical location of the online pharmacy. This is an important feature with regard to transparency, as well as to indicate their focal country/region (e.g., the USA, Canada, or Europe). (5) Stipulate the delivery criteria for purchases (i.e., from where to where). This information makes it easier for consumers to choose their online pharmacy based on the convenience of delivery, and whether it delivers to the desired location. Having said that, online pharmacies frequently deliver to all countries, including those in Asia. (6) Specify the prescription and OTC drugs available for purchase. Based on an analysis of the literature, the range of products sold by online pharmacies has become very complex over time; 10 years ago, they tended to generally sell lifestyle drugs, but these days they offer virtually any prescription or OTC drug. (7) Offer information on the drugs on sale, such as their indication, recommended dosages, administration routes, potential side effects (including ways to identify the symptoms and actions to be taken in such cases), and company/personnel contact details in case of inquiries and ambiguities. This information is also for the consumer's convenience. (8) Indicate the purchase price of drugs. This is one of the most important features of online pharmacies, as it allows consumers to compare online and retail pharmacy prices. (9) Have at least one quality-related certificate, which should accompany its medicine sale activities. It is a type of guarantee of the pharmacy's operational legitimacy, and the authenticity and quality of its products. It should ideally enable cross-checking with the FDA or at least the respective country's regulatory authorities. (10) Indicate how long they have been in business. (11) Provide their privacy policies and disclaimers.

16.5 Importance of Online Pharmacies

Distance/remote and online pharmacies have many benefits for patients and the public, especially those in rural areas or those who cannot access or reach the pharmacies and pharmacists. There are more benefits such as lower cost, convenience, greater anonymity for consumers, and saving time.

16.6 General Regulations of Online Pharmacies

Legal internet pharmacies must adhere to the laws and regulations of the specific country in which they are operating. The WHO, as well as the FDA, has identified the dangers of online pharmacies. Due to issues such as unapproved, fraudulent, and unsafe drugs being sold, the WHO initiated steps to tighten the control on the sale of drugs and medical products online (Kumaran et al., 2020).

16.7 Quality of Online Pharmacies

The quality of online pharmacies depends on their individual accreditation and verification, which is considered the most important criterion as follows (Kumaran et al., 2020; Gabay, 2015):

Licensure and Policy Maintenance

The pharmacy and pharmacists and pharmacy technicians are licensed; furthermore, the pharmacy should be licensed to provide online services and comply with all applicable statutes and regulations.

Prescriptions

Develop and enforce policies and procedures that ensure the integrity, legitimacy, and authority of the prescription drug order.

Prevent drug orders from being submitted and filled by multiple pharmacies.

Prevent medications from being prescribed or dispensed based upon telephonic or online medical consultations that do not result from a pre-existing prescriber-patient relationship.

Patient Information

Develop and enforce policies and procedures that ensure reasonable verification of the identity of the patient, prescriber, and caregiver if applicable.

Obtain and maintain patient medication profiles.

Conduct a prospective drug use review prior to medication dispensing.

Ensure patient confidentiality and protect patient-specific information when such information is transmitted over the internet.

Communication

Develop and enforce policies and procedures that require pharmacists to offer consultative services to patients.

Develop a system regarding reporting of adverse drug reactions and errors.

Develop a mechanism for contacting patients regarding delays in delivering a prescription medication as well as drug recalls.

Develop a mechanism for educating patients regarding disposal of expired, damaged, or unusable medications.

Storage and Shipment

Develop a system for shipping controlled substances safely and securely.

Ensure that medications are shipped to patients appropriately.

Over-the-Counter Products

Comply with all federal and state laws regarding the sale of over-the-counter products that may be used in the manufacture of illegal drugs.

Quality Improvement Programs

Maintain a quality improvement program.

16.8 Disadvantages of Online Pharmacies

There are many disadvantages and risks associated with buying medicines from online pharmacies such as counterfeit medication and low-quality medications, medications abuse and misuse, and self-medications with prescribed medications (Lee et al., 2017; Al-Worafi, 2020a–e). There are more risks related to the privacy of patients' and customers' information.

16.9 Conclusion

This chapter has discussed the history and importance of online pharmacies. This chapter includes background about online pharmacies, its definition, development, and importance. It also includes the disadvantages of online pharmacies and the regulations of online pharmacies.

References

Al-Ghazal, S.K. and Tekko, I.A., 2003. The valuable contributions of Al-Razi (Rhazes) in the history of pharmacy during the Middle Ages. *Journal of the International Society for the History of Islamic Medicine*, 2(9), pp. 9–11.

Al-Worafi, Y.M. ed., 2020a. *Drug safety in developing countries: Achievements and challenges*. Academic Press.

Al-Worafi, Y.M., 2020b. Counterfeit and substandard medications. In *Drug safety in developing countries* (pp. 119–126). Academic Press.

Al-Worafi, Y.M., 2020c. Drug safety in developing versus developed countries. In *Drug safety in developing countries* (pp. 613–615). Academic Press.

Al-Worafi, Y.M., 2020d. Medication abuse and misuse. In *Drug safety in developing countries* (pp. 127–135). Academic Press.

Al-Worafi, Y.M., 2020e. Self-medication. In *Drug safety in developing countries* (pp. 73–86). Academic Press.

Bender, G.A. and Thom, P.B.R.A., 1952. *A history of pharmacy in pictures*. Davis & Company.

Gabay, M., 2015. Regulation of internet pharmacies: A continuing challenge. *Hospital Pharmacy*, 50(8), p. 681.

Horgan, C., Goody, B., Knapp, D. and Fitterman, L., 1990. The role of mail service pharmacies. *Health Affairs*, 9(3), pp. 66–74

Kumaran, H., Long, C.S., Bakrin, F.S., Tan, C.S., Goh, K.W., Al-Worafi, Y.M., Lee, K.S., Lua, P.L. and Ming, L.C., 2020. Online pharmacies: Desirable characteristics and regulations. *Drugs & Therapy Perspectives*, 36(6), pp. 243–245.

Lee, K.S., Yee, S.M., Zaidi, S.T.R., Patel, R.P., Yang, Q., Al-Worafi, Y.M. and Ming, L.C., 2017. Combating sale of counterfeit and falsified medicines online: A losing battle. *Frontiers in Pharmacology*, 8, p. 268.

Online pharmacy, Wikipedia., 2021. Available at: https://en.wikipedia.org/wiki/Online_pharmacy

Nicole, Kruczek., 2020. *What is the role of the mail-order pharmacy?* Available at: www.pharmacytimes.com/view/what-is-the-role-of-the-mail-order-pharmacy

Orizio, G., Merla, A., Schulz, P.J. and Gelatti, U., 2011. Quality of online pharmacies and websites selling prescription drugs: A systematic review. *Journal of Medical Internet Research*, 13(3), p. e1795.

Rickert, E. and Anderson, D., 2000, March. Internet pharmacy practice: Legal and marketplace issues. In *Platform presentation.* at: Annual Meeting of the American Pharmaceutical Association.

17

Social Media, Social-Networking Sites, and Webinar and Video Conferencing Platforms

17.1 History of Social Media, Social-Networking Sites, and Webinar and Video Conferencing Platforms

The first recognizable social media site, Six Degrees, was created in 1997, followed by the first blogging sites in 1999 (Edosomwan et al., 2011). The impact of social media in pharmacy practice started after the evolution of modern social media. The history of modern social media is as follows:

Facebook: Launched in 2004

Twitter: Founded in 2006

Instagram: Founded in 2010 and purchased by Facebook in 2012

Pinterest: Founded in 2010 as a visual "pin board"

Snapchat: Founded in 2011

YouTube: Founded in 2005

WhatsApp: Founded in 2009 and purchased by Facebook in 2014

Wikipedia: Founded in 2001

Microsoft Teams: Founded in 2017

Cisco WebEx Teams: Founded in 2000

Google Meet: formerly known as Hangouts Meet, Founded in 2017

Zoom: Founded in 2011

DOI: 10.1201/9781003230458-19

17.2 Reasons for Using Social Media, Social-Networking Sites, and Webinar and Video Conferencing Platforms in Pharmacy Practice

There are many reasons for using social media, social-networking sites, and webinar and video conferencing platforms in pharmacy practice such as: Millions of the public, patients, and health care professionals in all countries around the world use social media on a daily basis; furthermore, all population groups such as adults, teenagers, elderlies, pregnant women, and nursing moms use social media on a daily basis which allows pharmacists and pharmacy technicians to use it to reach the public and patients for pharmacist care and patient care purposes. Pharmacists and pharmacy technicians can use it for training, workshops, conferences, and pharmacist care and patient care services. Pharmaceutical companies can use it for pharmaceutical marketing purposes. Pharmacists and pharmacy technicians can use it for public health programs.

17.3 Impact of Social Media and Social-Networking Sites in Pharmacy Practice

17.3.1 Facebook

Pharmacists and pharmacy technicians can use Facebook in many ways including the following: as a communication tool to communicate with patients, the public, and health care professionals. As an education tool to counsel and educate the public and patients about their health, diseases, medications, and others. Share experiences with colleagues in groups. Access profession related information and resources. Tool for conducting research and others. Pharmacists and pharmacy technicians can use it for training, workshops, conferences, and pharmacist care and patient care services. Pharmacists and pharmacy technicians can use it for public health programs.

17.3.2 Twitter

Pharmacists and pharmacy technicians can use Twitter in many ways including the following: As a communication tool to communicate with patients, the public, and health care professionals. As an education tool to counsel and educate the public and patients about their health, diseases, medications, and others. Share experiences and materials with colleagues in groups. Access profession related information and resources. Tool for conducting research and others. Pharmacists and pharmacy technicians can use it for training, workshops, conferences, and pharmacist care and patient care services. Pharmacists and pharmacy technicians can use it for public health programs.

17.3.3 Instagram

Pharmacists, pharmacy technicians can use Instagram in many ways including the following: As a communication tool to communicate with patients, the public, and health care professionals. As an education tool to counsel and educate the public and patients about their health, diseases, medications, and others. Share experiences and materials with colleagues in groups. Access profession related information and resources. Tool for conducting research and others. Share health information, discuss clinical cases, and listen to patient stories. Pharmacists and pharmacy technicians can use it for training, workshops, conferences, and pharmacist care and patient care services (Wong et al., 2019; Shafer et al., 2018; Gulati et al., 2020). Pharmacists and pharmacy technicians can use it for public health programs.

17.3.4 YouTube

YouTube can be used for streaming to the audience and having a one-way video interaction. YouTube can be a good resource for many pharmacy related education videos. Pharmacists and pharmacy technicians can use YouTube for recording lectures and sharing them with the public and patients on YouTube where they can easily access it at any time, download it if needed, as well as share it with their colleagues. Pharmacists and pharmacy technicians can use YouTube for training, workshops, conferences, and pharmacist care and patient care services. Pharmacists and pharmacy technicians can use it for public health programs.

17.3.5 WhatsApp

WhatsApp can be used in many ways including the following: communication tools; share educational resources; group chat; and others. Pharmacists and pharmacy technicians can use it for training, workshops, conferences, and pharmacist care and patient care services (Coleman et al., 2019; Raiman et al., 2017; us Salam et al., 2021). Pharmacists and pharmacy technicians can use it for public health programs.

17.3.6 Webinar and Video Conferencing Platforms

Webinar and video conferencing platforms are very important in online pharmacy practice. Microsoft Teams, Cisco WebEx Teams, Google Meet, and Zoom are the most common webinar and video conferencing platforms in online pharmacy education (www.microsoft.com/en-us/microsoft-teams/group-chat-software; www.webex.com/; https://apps.google.com/meet/; www.zoom.us/).

Microsoft Teams, Cisco WebEx Teams, Google Meet, and Zoom offer multiple versions of their software based on usage requirements. This includes free versions that are great for light uses, short conference calls, and light file sharing. However, pharmacists and pharmacy technicians can use it for training, workshops, conferences, and pharmacist care and patient care services. Pharmacists and pharmacy technicians can use it for public health programs.

17.4 Risks of Social Media

There are many potential risks for using social media such as (Ventola, 2014; George et al., 2013; Alsughayr, 2015): privacy and security issues, quality of information, and others.

17.5 Guidelines for Social Media Use by Pharmacists

The American Society of Health-System Pharmacists (ASHP) developed a statement about the use of social media by pharmacists and pharmacy institutions as follows: Participation in social media hospitals or health systems that choose to use social media or permit practice-related social media use by staff should have in place policies and procedures that: Balance the benefits social media provide with the obligations and liabilities they may create; encourage the development and application of best practices by users of social media. Pharmacy professionals are encouraged to exercise their professional judgment in incorporating social media into their practices. Advancing the well-being and dignity of patients. The following recommendations can help pharmacy professionals who choose to participate in social media to advance the well-being and dignity of patients. Acting with integrity and conscience. The following recommendations are intended to assist pharmacy professionals to act with integrity and conscience in their use of social media. Collaborating respectfully with health care colleagues. Although social media can and should be used to promote healthy debate about health care and pharmacy practice, such debate should be conducted in a respectful manner. Reasoned debate sometimes requires constructive criticism, but pharmacy professionals should not use social media to make ad hominem comments or needlessly denigrate specific care providers, institutions, or professions. Pharmacy professionals should continue to adhere to all laws, regulations, standards, and other mandates intended to protect patient privacy and confidentiality in social media.

17.6 Conclusion

This chapter has discussed the history and importance of social media, social-networking sites, and webinar and video conferencing. It includes the impact of social media, social-networking sites, and webinar and video conferencing and the important role they play in pharmacy practice and patient care.

References

Alsughayr, A.R., 2015. Social media in healthcare: Uses, risks, and barriers. *Saudi Journal of Medicine and Medical Sciences*, 3(2), p. 105.

American Society of Health-System Pharmacists (ASHP)., 2012. ASHP statement on use of social media by pharmacy professionals: Developed through the ASHP pharmacy student forum and the ASHP section of pharmacy informatics and technology and approved by the ASHP Board of Directors on April 13, 2012, and by the ASHP House of Delegates on June 10, 2012. *American Journal of Health-System Pharmacy: AJHP: Official Journal of the American Society of Health-System Pharmacists*, 69(23), pp. 2095–2097. Available at: www.ashp.org/-/media/assets/policy-guidelines/docs/statements/use-of-social-media-by-pharmacy-professionals.ashx

Coleman, E. and O'Connor, E., 2019. The role of WhatsApp® in medical education; a scoping review and instructional design model. *BMC Medical Education*, 19(1), pp. 1–13.

Edosomwan, S., Prakasan, S.K., Kouame, D., Watson, J. and Seymour, T., 2011. The history of social media and its impact on business. *Journal of Applied Management and Entrepreneurship*, 16(3), pp. 79–91.

George, D.R., Rovniak, L.S. and Kraschnewski, J.L., 2013. Dangers and opportunities for social media in medicine. *Clinical Obstetrics and Gynecology*, 56(3).

Gulati, R.R., Reid, H. and Gill, M., 2020. Instagram for peer teaching: opportunity and challenge. *Education for Primary Care*, 31(6), pp. 382–384.

Raiman, L., Antbring, R. and Mahmood, A., 2017. WhatsApp messenger as a tool to supplement medical education for medical students on clinical attachment. *BMC Medical Education*, 17(1), pp. 1–9.

Shafer, S., Johnson, M.B., Thomas, R.B., Johnson, P.T. and Fishman, E.K., 2018. Instagram as a vehicle for education: What radiology educators need to know. *Academic Radiology*, 25(6), pp. 819–822.

us Salam, M.A., Oyekwe, G.C., Ghani, S.A. and Choudhury, R.I., 2021. How can WhatsApp® facilitate the future of medical education and clinical practice? *BMC medical Education*, 21(1), pp. 1–4.

Ventola, C.L., 2014. Social media and health care professionals: benefits, risks, and best practices. *Pharmacy and Therapeutics*, 39(7), p. 491.

Wong, X.L., Liu, R.C. and Sebaratnam, D.F., 2019. Evolving role of Instagram in# medicine. *Internal Medicine Journal*, 49(10), pp. 1329–1332.

18

Mobile Health Technologies

18.1 Terminologies

A mobile device is defined as a handheld computing device characterized by a touch-screen display for input, streamlined operating system, and apps (Hanrahan et al., 2014).

Mobile Health (m-Health or mhealth): The World Health Organization (WHO), in collaboration with the Global Observatory for eHealth, has defined m-Health as "medical and public health practice supported by mobile devices, such as mobile phones, patient monitoring devices, personal digital assistants (PDAs), and other wireless devices" (WHO, 2011).

Mobile application (app): Software for use on a mobile platform. Downloadable mobile app: Mobile apps that are developed by third-party organizations and then downloaded to the device hardware (Hanrahan et al., 2014).

Smartphone: Small portable mobile device focused on communication through messaging and voice-to-voice communication supported by cellular services (Hanrahan et al., 2014).

Short messaging service (SMS): Services dedicated towards communication through messages consisting of text (i.e., texting) (Hanrahan et al., 2014).

Social media: Networks of communication focused on electronic interactions between users where content is shared, created, and ideas exchanged (Hanrahan et al., 2014).

18.2 History of Mobile Health (m-Health or mhealth)

The history of Mobile health goes back to the 2000s, when mobile devices were developed and became popular around the world. Mobile Health (m-Health or mhealth) started with short messages through mobiles devices. Smartphones have played an important role in the development of Mobile

Health (m-Health or mhealth) since 2009. The following are examples of Mobile Health applications history (Pai et al., 2013):

February 6, 2009: iTMP announces the launch of its Smheart Link device, a "wireless bridge for biometrics" that connects off-the-shelf fitness sensors like heart rate straps to a user's iPhone.

April 7, 2010: Entra's MyGlucoHealth Diabetes App launched.

September 21, 2010: French pharmaceutical company Sanofi Aventis announces that it has tapped medical device maker Agamatrix to create blood glucose meter plug-in for Apple's iPhone called iBGStar.

January 4, 2011: iHealth Lab, a San Francisco-based subsidiary of Chinese medical company Andon Health, announces the iHealth Blood Pressure Monitoring System for iPhone.

January 20, 2011: The FDA clears the Withings iPhone blood pressure cuff as a Class II medical device.

January 13, 2012: UK-based retail pharmacy chain Lloyds pharmacy inks a deal with Proteus Biomedical to launch Proteus' first commercial product, Helius, an offering that includes sensor-enabled pills, a peel-and-stick sensor patch worn on the body, and a mobile health app.

January 30, 2012: Glooko offers a simple glucose monitoring logbook app and a cable that connects meters to iPhones.

18.3 Medical Uses of Mobile Health (m-Health or mhealth)

Mobile Health (m-Health or mhealth) can be used in the following (Adibi, 2014; Consulting, 2009):

- Education and awareness
- Helpline
- Diagnostic and treatment support
- Communication and training for health care workers
- Disease and epidemic outbreak tracking
- Remote monitoring
- Remote data collection

18.4 Reasons for Using Mobile Health (m-Health or mhealth) in Pharmacy Practice

There are many reasons for using Mobile Health (m-Health or mhealth) in pharmacy practice such as: Millions of the public, patients, and health care professionals in all countries around the world use Mobile Health (m-Health or mhealth) on a daily basis; furthermore, all population groups such as adults, teenagers, elderlies, pregnant women, and nursing moms use Mobile Health (m-Health or mhealth) on a daily basis which allows pharmacists and pharmacy technicians to use it to reach the public and patients for pharmacist care and patient care purposes.

18.5 Impact of Mobile Health (m-Health or mhealth) in Pharmacy Practice

Pharmacists and pharmacy technicians can use Mobile Health (m-Health or mhealth) to enable and support pharmacists within their pharmacist care and patient care roles as follows (FIP, 2019; Ming et al., 2016; Ming et al., 2020; Izahar et al., 2017; AL-Worafi, 2020a, b):

Drug Information Resources

The availability of drug/medication reference texts and websites on mobile devices helps pharmacists have access to trusted and reliable references at the point of patient care. This supports productivity and sound clinical decision making. Examples of reference categories and functionality on mobile devices are described next.

Drug Information

Clinical reference texts and drug databases are increasingly available via mobile devices, including searches for drug indications, dosages, contraindications, interactions, adverse drug reactions, availability, and other information.

Examples: Micromedex, Lexicomp, Medscape, MIMS, BNF and others.

Calculators

Clinical calculators are intended to guide clinical decision making, such as for drug dosing of drugs with a narrow therapeutic index such as aminoglycosides, vancomycin, or phenytoin dosing, or clinical indicators such as creatinine clearance.

Examples: QxMD, Lexicomp, ClinCalc Medical Calculator, and others.

Guidelines

Guidelines provide evidence-based recommendations to pharmacists. These tools are used in a pharmacist's daily practice to provide the best patient care possible.

Examples: International Society for Peritoneal Dialysis guidelines, oncology guidelines available through the US National Comprehensive Cancer Network and Epocrates, Sanford Guide for the Antimicrobial Therapy, American College of Cardiology/American Heart Association for hypertension guidelines, and others.

Literature Databases

Mobile devices can often link to academic databases, including health journals, where internet access is possible. This functionality, through web browsers or apps, can provide pharmacists with access to resources.

Examples: British Medical Journal, New England Journal of Medicine, The Lancet, Pubmed, MEDline.

Continuing Education and Professional Development Activities

Continuing education and professional development are a part of the duties of a practicing pharmacist in order to stay up to date on the latest medical treatments and services. Mobile devices provide an accessible platform for pharmacists to perform these activities. Tools that support continuing education provide pharmacists with patient case studies, lessons on new treatment options, treatment reviews, and more.

Examples: Medscape Education, BMJ Best Practice, Pharmacy Times, Online Academy.

Diagnostic Support Tools/Point-of-Care Diagnostics

Some point-of-care diagnostic therapeutic devices are now designed to be used with mobile devices, such as those that measure respiratory function or blood glucose levels. Although these are primarily intended for consumer use in the management of chronic health conditions (such as diabetes), their connectivity and convenience mean they can be used by pharmacists in screening or monitoring services.

Examples: Dexcom Continuous Glucose Monitoring, Air Smart Spirometer, KardiaMobile.

Medicines Availability

There are mobile tools which provide information to pharmacists about medicines availability, shortages, and alternative treatments. Drug shortages affect daily practice and it is important for pharmacists to be able to access this information easily and quickly.

Examples: Drug Shortages, Orange Book, Food Safety Alerts & Tips, FDA Recalls, Market Withdrawals & Safety Alerts.

Patient Information Repository

Some apps provide pharmacists with patient health information and patient prescription histories. Many institutions have their own specific app which contains patient medical records. These apps allow pharmacists to have access to the patient's medical information when computers are not available or convenient to use.

Examples: Epic Haiku, MySNS Wallet, Patient Portal, Care360 Mobile.

Communication Tools

Communication tools can be used to communicate with patients, the public, and health care professionals. They can also be used as education tools to counsel and educate the public and patients about their health, diseases, medications, and others. They can be used to share experiences with colleagues in groups, to access profession related information and resources, and to conduct research.

Tools for Social Media, Network, Webinar, and Video Conferencing Platforms

Mobile devices and smartphones can be used by pharmacists and pharmacy technicians to access and use Facebook, Twitter, Instagram, YouTube, WhatsApp, webinar and video conferencing platforms. These tools are very important in online pharmacy practice using Microsoft Teams, Cisco WebEx Teams, Google Meet, and Zoom for pharmacist care and patient care purposes.

Medication Safety

Pharmacists and pharmacy technicians have used and can use mobile devices, smartphones, and mobile apps for medication safety related issues such as pharmacovigilance, adverse drug reactions (ADRs) reporting, medication errors reporting, and others.

18.6 Mobile Health (m-Health or mhealth) Regulations

Historically, software products intended for use in the diagnosis or treatment of disease have been classified as a medical device by the U.S. Food and Drug Administration (FDA). The regulation of medical devices differs from that of drugs since it is based on a three-tier classification system. Specifically, devices are designated as either Class I, II, or III, depending on their potential risk. *Class I devices* are considered to be the lowest risk and are generally exempt from FDA review. *Class II devices*, however, are considered an intermediate level of risk and developers are usually required to submit a premarket notification. *Class III devices* are considered to be the highest risk level and must generally undergo a more complex, time-consuming, and expensive premarket approval process. The regulation of medical apps was not specifically addressed until 2011, when the FDA released a draft guidance on the topic. The guidance, which was updated and finalized in September 2013, outlines how the FDA will apply its regulatory authority to mobile medical apps (FDA, 2013) as follows:

Regulated Apps

- Control other medical devices.
- Display, store, analyze, or transmit patient-specific medical data from another device.
- Use attachments, display screens, or sensors to transform the mobile platform into a medical device.
- Perform patient-specific analysis and provide a patient-specific diagnosis or treatment recommendation.

Apps Subject to Enforcement Discretion

- Provide or facilitate supplemental care by coaching or prompting patients.
- Help patients organize or track health information.
- Provide access to information related to health conditions or treatments.
- Allow patients to communicate medical conditions with providers.
- Perform simple calculations used in clinical practice.
- Enable individuals to interact with electronic health records.

Apps That Will Not Be Regulated

- Include electronic copies of medical textbooks, teaching aids, or other reference materials.
- Are intended as educational tools for medical training.
- Facilitate patient access or understanding.
- Automate general office operations.
- Are not specifically designed or intended for medical purposes.

Quality of Mobile Health (m-Health or mhealth)

There is a need to develop and validate the evaluation checklists for medical apps, to ensure the quality of apps.

18.9 Conclusion

This chapter has discussed the mobile health technologies and their impact on pharmacy practice. There are many reasons for using Mobile Health (m-Health or mhealth) in pharmacy practice such as: Millions of the public, patients, and health care professionals in all countries around the world use Mobile Health (m-Health or mhealth) on daily basis; furthermore, all population groups such as adults, teenagers, elderlies, pregnant women, and nursing moms use Mobile Health (m-Health or mhealth) on a daily basis which allows pharmacists and pharmacy technicians to use it to reach the public and patients for pharmacist care and patient care purposes.

References

Adibi, S. ed., 2014. *mHealth multidisciplinary verticals*. CRC Press.

Al-Worafi, Y.M., 2020a. Technology in medications safety. In *Drug safety in developing countries* (pp. 203–212). Academic Press.

Al-Worafi, Y.M., 2020b. Drug safety in developing versus developed countries. In *Drug safety in developing countries* (pp. 613–615). Academic Press.

Consulting, V.W., 2009. *mHealth for development: the opportunity of mobile technology for healthcare in the developing world*. Available at: http://www.globalproblems-globalsolutions-files.org/unf_website/assets/publications/technology/mhealth/mHealth_for_Development_full.pdf

FDA., 2013. *Mobile medical applications: Guidance for industry and food and drug administration staff* [Internet]. US Food and Drug Administration. Available at: www.fda.gov/downloads/MedicalDevices/DeviceRegulationandGuidance/GuidanceDocuments/UCM263366.pdf. www.fda.gov/MedicalDevices/DeviceRegulationandGuidance/overview/.

Hanrahan, C., Aungst, T.D. and Cole, S., 2014. Evaluating mobile medical applications. *American Society of Health-System Pharmacists*. Available at: https://www.ashp.org/-/media/store%20files/mobile-medical-apps.pdf

International Pharmaceutical Federation (FIP)., 2019. *mHealth: Use of mobile health tools in pharmacy practice*. International Pharmaceutical Federation.

Izahar, S., Lean, Q.Y., Hameed, M.A., Murugiah, M.K., Patel, R.P., Al-Worafi, Y.M., Wong, T.W. and Ming, L.C., 2017. Content analysis of mobile health applications on diabetes mellitus. *Frontiers in Endocrinology*, 8, p. 318.

Ming, L.C., Hameed, M.A., Lee, D.D., Apidi, N.A., Lai, P.S.M., Hadi, M.A., Al-Worafi, Y.M.A. and Khan, T.M., 2016. Use of medical mobile applications among hospital pharmacists in Malaysia. *Therapeutic Innovation & Regulatory Science*, 50(4), pp. 419–426.

Ming, L.C., Untong, N., Aliudin, N.A., Osili, N., Kifli, N., Tan, C.S., Goh, K.W., Ng, P.W., Al-Worafi, Y.M., Lee, K.S. and Goh, H.P., 2020. Mobile health apps on COVID-19 launched in the early days of the pandemic: Content analysis and review. *JMIR mHealth and uHealth*, 8(9), p. e19796.

Pai, A., Comstock, J. and Dolan, B., 2013. Timeline: Smartphone-enabled health devices. *MobiHealthNews* (7 June 2013).

World Health Organization., 2011. *mHealth: New horizons for health through mobile technologies*. mHealth: New Horizons for Health through Mobile Technologies.

19

Medications Safety

19.1 Background

Pharmacists nowadays contribute effectively to medication safety practices in developed countries and in many developing countries. Pharmacists play an important role in the activities and monitoring related to the effectiveness, quality, and safety of medications, herbal medications, vaccinations, and other medicinal products which include: medication registration (licensing)/re-registration (re-licensing); pharmacovigilance (for medications, herbal medications, vaccinations, self-medications, and medications abuse and misuse), and adverse drug reactions (ADRs) and its reporting; medication errors and its reporting; drug related problems (DRPs); counterfeit medications; storage and disposal of medications; rationality and appropriate use of medications such as antibiotics; and other medication safety concerns. The new technologies/information technologies have contributed effectively to medication health care during the last decades. Adapting new technologies/ information technologies, mobile technologies, and social media has contributed/could contribute effectively to medications safety practice such as pharmacovigilance, medication errors, drug-related problems (DRPs), and other safety practices; however, on the other hand lack, of technologies is a major barrier for effective medication safety practice (Huckvale et al., 2010; Bates and Gawande, 2003; Bates, 2000; Ming et al., 2016; Izahar et al., 2017; Al-Worafi, 2020a–j; Kaushal and Bates, 2002; Chaudhry et al., 2006; Lee et al., 2017; Christy, 2016; Elangovan et al., 2020).

19.2 Online Pharmacy Practice: Pharmacovigilance, Adverse Drug Reactions (ADRs) and Its Reporting

Pharmacovigilance (PV) is defined as the science and activities relating to the detection, assessment, understanding, and prevention of adverse effects or any other drug-related problem (WHO, 2002). The American

Society of Health-System Pharmacists (ASHP, 1995) defined ADRs as: "Any unexpected, unintended, undesired, or excessive response to a drug that requires discontinuing the drug (therapeutic or diagnostic), requires changing the drug therapy, requires modifying the dose (except for minor dosage adjustments), necessitates admission to a hospital, prolongs stay in a health care facility, necessitates supportive treatment, significantly complicates diagnosis, negatively affects prognosis, or results in temporary or permanent harm, disability, or death" (ASHP, 1995). Adverse drug reactions (ADRs) are common worldwide and are associated with more morbidity and mortality, and worse clinical, economic, and humanistic outcomes (Al-Worafi, 2020a–d, 2020k–o; Al-Worafi et al., 2017; Elsayed and Al-Worafi., 2020; Elkalmi et al., 2020; Baig et al., 2020; Hasan et al., 2019). Pharmacists play an important role in the detection, reporting, prevention, and management of adverse drug reactions (ADRs). Pharmacists have used/can use new technologies such as mobile applications, social media, and other technologies for success in online medication safety practice. Pharmacists can encourage patients to communicate with them through WhatsApp, social media, and other technologies to report any ADR, and to counsel and educate patients about it. New technologies such as mobile applications can be used as easy, accessible, and free tools to submit ADR reports to the national pharmacovigilance centers.

19.3 Online Pharmacy Practice: Medication Errors and Its Reporting

Medication errors is defined as "episodes in drug misadventuring that should be preventable through effective systems controls involving pharmacists, physicians and other prescribers, nurses, risk management personnel, legal counsel, administrators, patients, and others in the organizational setting, as well as regulatory agencies and the pharmaceutical industry" (ASHP, 1993). Prescribing errors is defined as any error related to identification of patient-related problems; gathering patient-related information; medical and medications histories; assessment; management plan which includes objective and desired outcomes; nonpharmacological recommendations such as weight control; appropriate and rational pharmacological recommendations with doses; dosage from route of administration, frequency, and duration; time of taking medications and instructions; monitoring for efficacy and safety, as well as disease; patient education and counseling related to adherence towards the management plan, self-management, potential adverse drug effects and reactions, possible interactions, caution and precautions, contraindications and warnings, proper storage and disposal of medications (Al-Worafi, 2020). Prescription writing errors is defined as the type of error that occurs when the prescription elements are either not written or are written incorrectly. These include the following related errors: (1) errors related to physician or authorized prescriber—name, contact details, and signature; (2) errors related to patient information—name, address, age, gender, and weight; (3) errors related to prescribed medications—drug name, strength, dose units, dosage form, quantity of medications, duration of therapy, route of administration, dose interval, instructions, drug abbreviation, unit abbreviation, spelling; and (4) errors related to the whole prescription such as date of prescription, diagnosis, and clarity of prescription (Al-Worafi et al., 2018). Dispensing errors is defined as errors that occur when distributing or selling prescription to patients or patients' agents (Flynn et al., 2003). Medications errors such as prescribing errors, prescription writing errors, and dispensing errors are common worldwide and associated with more morbidity and mortality, and worse clinical, economic, and humanistic outcomes (Al-Worafi, 2020a–d, 2020k–o; Al-Worafi et al., 2017; Elsayed and Al-Worafi, 2020; Elkalmi et al., 2020; Baig et al., 2020). Pharmacists play an important role in the detection, reporting, and prevention and management of medication errors. Pharmacists have used/can use new technologies such as mobile applications, social media, and other technologies for success in online medication safety practice. Pharmacists can encourage patients to communicate with them through WhatsApp, social media, and other technologies to report any medication errors, and to counsel and educate patients about it. New technologies such as mobile applications can be used as easy, accessible, and free tools to submit medication errors reports to the national centers.

19.4 Online Pharmacy Practice: Medication Misuse and Abuse

Pharmacists can educate and counsel patients about potential medications abuse and misuse through WhatsApp, social media, and other technologies.

19.5 Online Pharmacy Practice: Self-Medications

Pharmacists can educate and counsel patients about the appropriate use of self-medications through WhatsApp, social media, and other technologies.

19.6 Online Pharmacy Practice: Substandard and Counterfeit Medications

Pharmacists play a very important role in fighting substandard and counterfeit medications. Pharmacists can encourage patients to communicate with them through WhatsApp, social media, and other technologies to report any medication errors, and to counsel and educate patients about the importance of buying and using licensed medications and herbal medications and nutraceuticals, as the substandard and counterfeit medications will affect their health and cause many dangerous complications and effects. New technologies such as mobile applications can be used as easy, accessible, and free tools to submit reports about substandard and counterfeit medications to the national pharmacovigilance centers.

19.7 Conclusion

This chapter has discussed online medications safety practices. The new technologies/information technologies have contributed effectively to medication health care during the last decades, adapting new technologies/information technologies, mobile technologies, and social media has contributed/could contribute effectively to medications safety practice such as pharmacovigilance, medication errors, drug related problems (DRPs), and other safety practices.

References

American Society of Health-System Pharmacists (ASHP), 1993. Guidelines on preventing medication errors in hospitals. *American Journal of Health-System Pharmacy*, 50, pp. 305–314.

American Society of Health-System Pharmacists (ASHP), 1995. ASHP guidelines on adverse drug reaction monitoring and reporting: American society of hospital pharmacy. *American Journal of Health-System Pharmacy*, 52(4), pp. 417–419.

Al-Worafi, Y.M., 2020a. Technology in medications safety. In *Drug safety in developing countries* (pp. 203–212). Academic Press.

Al-Worafi, Y.M., 2020b. Drug safety in developing versus developed countries. In *Drug safety in developing countries* (pp. 613–615). Academic Press.

Al-Worafi, Y.M., 2020c. Pharmacovigilance. In *Drug safety in developing countries* (pp. 29–38). Academic Press.

Al-Worafi, Y.M., 2020d. Adverse drug reactions. In *Drug safety in developing countries* (pp. 39–57). Academic Press.

Al-Worafi, Y.M., 2020e. Self-medication. In *Drug safety in developing countries* (pp. 73–86). Academic Press.

Al-Worafi, Y.M., 2020f. Medication errors. In *Drug safety in developing countries* (pp. 59–71). Academic Press.

Al-Worafi, Y.M., 2020g. Counterfeit and substandard medications. In *Drug safety in developing countries* (pp. 119–126). Academic Press.

Al-Worafi, Y.M., 2020h. Medication abuse and misuse. In *Drug safety in developing countries* (pp. 127–135). Academic Press.

Al-Worafi, Y.M., 2020i. Drug-related problems. In *Drug safety in developing countries* (pp. 105–117). Academic Press.

Al-Worafi, Y.M., 2020j. Antibiotics safety issues. In *Drug safety in developing countries* (pp. 87–103). Academic Press.

Al-Worafi, Y.M., 2020k. Drug safety in Yemen. In *Drug safety in developing countries* (pp. 391–405). Academic Press.

Al-Worafi, Y.M., Patel, R.P., Zaidi, S.T., Alseragi, W.M., Almutairi, M.S., Alkhoshaiban, A.S. and Ming, L.C., 2018. Completeness and legibility of handwritten prescriptions in Sana'a, Yemen. *Medical Principles and Practice*, 27, pp. 290–292.

Baig, M.R., Al-Worafi, Y.M., Alseragi, W.M., Ming, L.C. and Siddique, A., 2020. Drug safety in India. In *Drug safety in developing countries* (pp. 327–334). Academic Press.

Al-Worafi, Y.M., 2020L. Drug safety in Saudi Arabia. In *Drug safety in developing countries* (pp. 407–417). Academic Press.

Al-Worafi, Y.M., 2020m. Drug safety in United Arab Emirates. In *Drug safety in developing countries* (pp. 419–428). Academic Press.

Al-Worafi, Y.M., 2020n. Drug safety in Indonesia. In *Drug safety in developing countries* (pp. 279–285). Academic Press.

Al-Worafi, Y.M., 2020o. Drug safety in Palestine. In *Drug safety in developing countries* (pp. 471–480). Academic Press.

Al-Worafi, Y.M., 2020p. Medications safety-related terminology. In *Drug safety in developing countries* (pp. 7–19). Academic Press.

Al-Worafi, Y.M., Alseragi, W.M., Ming, L.C. and Alakhali, K.M., 2020. Drug safety in China. In *Drug safety in developing countries* (pp. 381–388). Academic Press.

Al-Worafi, Y.M., Kassab, Y.W., Alseragi, W.M., Almutairi, M.S., Ahmed, A., Ming, L.C., Alkhoshaiban, A.S. and Hadi, M.A., 2017. Pharmacovigilance and adverse drug reaction reporting: A perspective of community pharmacists and pharmacy technicians in Sana'a, Yemen. *Therapeutics and Clinical Risk Management*, 13, p. 1175.

Al-Worafi, Y.M., Patel, R.P., Zaidi, S.T.R., Alseragi, W.M., Almutairi, M.S., Alkhoshaiban, A.S. and Ming, L.C., 2018. Completeness and legibility of handwritten prescriptions in Sana'a, Yemen. *Medical Principles and Practice*, 27, pp. 290–292.

Bates, D.W., 2000. Using information technology to reduce rates of medication errors in hospitals. *BMJ*, 320(7237), pp. 788–791.

Bates, D.W. and Gawande, A.A., 2003. Improving safety with information technology. *New England Journal of Medicine*, 348(25), pp. 2526–2534.

Chaudhry, B., Wang, J., Wu, S., Maglione, M., Mojica, W., Roth, E., Morton, S.C. and Shekelle, P.G., 2006. Systematic review: impact of health information technology on quality, efficiency, and costs of medical care. *Annals of Internal Medicine*, 144(10), pp. 742–752

Christy Wilson., 2016 *New technologies are accelerating drug development, bringing hope to patients*. Available at: www.elsevier.com/connect/new-technologies-are-accelerating-drug-development-bringing-hope-to-patients

Elangovan, D., Long, C.S., Bakrin, F.S., Tan, C.S., Goh, K.W., Hussain, Z., Al-Worafi, Y.M., Lee, K.S., Kassab, Y.W. and Ming, L.C., 2020. Application of blockchain technology in hospital information system. *Mathematical Modeling and Soft Computing in Epidemiology*, pp. 231–246.

Elkalmi, R.M., Al-Worafi, Y.M., Alseragi, W.M., Ming, L.C. and Siddique, A., 2020. Drug safety in Malaysia. In *Drug safety in developing countries* (pp. 245–253). Academic Press.

Elsayed, T. and Al-Worafi, Y.M., 2020. Drug safety in Egypt. In *Drug safety in developing countries* (pp. 511–523). Academic Press.

Flynn, E.A., Barker, K.N. and Carnahan, B.J., 2003. National observational study of prescription dispensing accuracy and safety in 50 pharmacies. *Journal of the American Pharmaceutical Association* (1996), 43(2), pp. 191–200.

Hasan, S., Al-Omar, M.J., AlZubaidy, H. and Al-Worafi, Y.M., 2019. Use of medications in Arab Countries. *Handbook of healthcare in the Arab world*. Springer, 42.

Huckvale, C., Car, J., Akiyama, M., Jaafar, S., Khoja, T., Khalid, A.B., Sheikh, A. and Majeed, A., 2010. Information technology for patient safety. *BMJ Quality & Safety*, 19(Suppl 2), pp. i25–i33.

Izahar, S., Lean, Q.Y., Hameed, M.A., Murugiah, M.K., Patel, R.P., Al-Worafi, Y.M., Wong, T.W. and Ming, L.C., 2017. Content analysis of mobile health applications on diabetes mellitus. *Frontiers in Endocrinology*, 8, p. 318.

Kaushal, R. and Bates, D.W., 2002. Information technology and medication safety: What is the benefit? *BMJ Quality & Safety*, 11(3), pp. 261–265.

Lee, K.S., Yee, S.M., Zaidi, S.T.R., Patel, R.P., Yang, Q., Al-Worafi, Y.M. and Ming, L.C., 2017. Combating sale of counterfeit and falsified medicines online: A losing battle. *Frontiers in Pharmacology*, 8, p. 268.

Ming, L.C., Hameed, M.A., Lee, D.D., Apidi, N.A., Lai, P.S.M., Hadi, M.A., Al-Worafi, Y.M.A. and Khan, T.M., 2016. Use of medical mobile applications among hospital pharmacists in Malaysia. *Therapeutic Innovation & Regulatory Science*, 50(4), pp. 419–426.

World Health Organization (WHO), 2002. *The importance of pharmacovigilance*. WHO, Geneva.

20

Patient Care

20.1 Background

Pharmacists nowadays play an important role in the health care system. Pharmacist nowadays play/should play an important role in patient care, as they contribute effectively to patients' health, diseases/conditions management as well as prevention; treating outcomes; treating cost; patients' quality of life; satisfaction towards the health care system and care. Pharmacists' roles and responsibilities and contributions have changed from a focus on the products (medicines and others) towards more advanced patient and public care. The new technologies/information technologies have contributed effectively to medication health care during the last decades, adapting new technologies/information technologies, mobile technologies, and social media has contributed/could contribute effectively to patient care, and facilitate health care services, especially in rural areas. During 2020 and 2021 the majority of outpatient clinics were closed in many countries around the world, which lead to an increase in the demand for e-health. New technologies played a very important role to facilitate e-health. Pharmacists during the lockdown and forwards used the new technologies to provide effective online pharmacist care and patient care services such as dispensing medications and herbal medicines and nutraceuticals; medications review; medication therapy management (MTM); patient education and counseling, and others.

20.2 Online Patient Care: Dispensing Medications

Pharmacists can dispense OTC medications, herbal medications, and nutraceuticals; furthermore, pharmacists can dispense prescribed medications online with valid online prescriptions and orders from the registered prescribers as well as by prescribed pharmacists. Many pharmacies in many countries around the world are licensed as online pharmacies also, which helps the pharmacists to provide effective online dispensing services with the help of new technologies.

20.3 Online Patient Care: Patient Assessment

Pharmacists can interview patients online by using Cisco WebEx Teams, Google Meet, and Zoom or others. Pharmacists can first gather the patients' related information: age, gender, weight, height, Body Mass Index (BMI); chief complaint, and history of present illness; medical history; medications history, family history, surgical history, and social history and document it. Pharmacists can ask patients to measure/estimate their vital signs (if applicable) such as: BP, Temp, RR, HR, and so on and document it. New technologies such as mobile applications can help pharmacists and patients in patient assessment. There are many mobile/wearable applications that can help in patient assessment; otherwise, pharmacists can ask the patients to measure vital signs in addition to the laboratory and diagnostic tests and requirements (if needed) at the nearest health care facility and send the results to pharmacists. This step is very important in patient care, and it will help pharmacists to design effective management plans for the patients.

20.4 Online Patient Care: Identify Patient Needs and Drug-Related Problems (DRPs)

Pharmacists can identify the patient's needs and actual and potential drug-related problems (DRPs) online by gathering the patient's related information and patient assessment as explained in the previous section, then design a management plan for patients including solving actual DRPs and preventing/minimizing potential DRPs.

20.5 Online Patient Care: Management Plan

Pharmacists can design and implement the management plan for their patients online. The management plan should include the following (Al-Worafi, 2020):

Goals of therapy and desired outcomes for all diseases/conditions.

Nonpharmacological therapies (individualize the "life style changes" such as weight control, healthy dietary therapy, increasing physical activity, modifying the modifiable risk factors, and so on depending on the disease and patient situation).

Pharmacological therapies (appropriate and rational based on the guidelines and recommendations with doses, dosage form and route of administration, strength, frequency, duration; time of taking medications and instructions).

Monitoring parameters:

The efficacy of medications (Is the prescribed medication effective? Is the desired outcome achieved?). This can be done by using the laboratory results, checking the symptoms improvement, patients' report, and other criteria.

The safety of medications (Is the prescribed medications safe?). This can be done by patients' reports about side/adverse effects/reactions, evaluating the effects on the patient's different systems such as renal, liver, and so on, and requesting laboratory tests, requesting drug therapy monitoring (TDM), and others.

Adherence towards the management plan.

Therapy success and complications: Is the treatment desired outcome achieved?

Patient education and counseling related to adherence towards the management plan (nonpharmacological, pharmacological therapies, and monitoring parameters), self-management, potential adverse drug effects and reactions, possible interactions, cautions and precautions, contraindications and warnings, proper storage and disposal of medications.

20.6 Online Patient Care: Medication Review

Medication Review or Medication Use Evaluation or Medication Use Review or Drug Utilization Review can be defined as a systematic and interdisciplinary performance improvement method with an overarching goal of optimizing patient outcomes via ongoing evaluation and improvement of medication utilization. OR "structured evaluation of patient's medicines with the aim of optimizing medicines use and improving health outcomes. This entails detecting drug related problems and recommending interventions" OR a patient-centered approach to optimize medication use and improve patient outcomes by ensuring each patient's medication is assessed for indication, effectiveness, and safety given patient status and comorbidities (ASHP, 1996; Kubacka, 1996; Afanasjeva, 2021; WHO, 2003; FIP, 2020; CMM, 2018). Pharmacists can use new technologies and perform the Medication Review as follows (FIP, 2020):

20.6.1 Collecting All Necessary Data with the Patient's Consent, If Required

20.6.1.1 Reviewing Diagnoses and Medication

- Is each medicine still indicated?
- Is each diagnosed disease being treated by a medicine?
- If the patient has renal or hepatic impairment, do medication dosages require adjusting?
- For each medicine, are there any adverse effects or laboratory markers to follow?
- Could there be there any medicine-medicine, medicine-disease, or medicine-natural product interactions?
- Can the dosing regimen or route of administration be simplified?
- Are there any more cost-effective alternatives to this medicine? Have new guidelines
- reinforced or discouraged its use?
- Are there any nonpharmacological methods that may be used?
- Do any natural health products or complementary or traditional medicines require intervention?

20.6.2 Reviewing Patient's Level of Literacy and Self-Monitoring

- Does the patient understand their medicines and their indications?
- Is the patient capable of self-monitoring, if required (blood glucose, blood pressure, etc.)?
- Is the patient aware of red flag symptoms that would require an urgent medical consultation?

20.6.3 Reassessing Medicines Management and Adherence

- Are medicine formulations and dosing schedules convenient for the patient?
- Can medication management be improved, for example, through the use of pillboxes?
- Is the patient adherent to their dosing schedules?

20.6.4 Organizing Follow-Up Visits to Assess Symptoms, Laboratory Markers, or Other Features to Monitor

20.6.5 Communicating with Prescribers and Other Allied Health Care Professionals regarding Suggested Changes, and Informing the Patient of the Results

20.7 Online Patient Care: Medication Therapy Management (MTM)

Medication therapy management (MTM) is a distinct service or group of services provided by health care providers, including pharmacists, to ensure the best therapeutic outcomes for patients. Pharmacists, with the help of new technologies, can perform effective online MTM with the core elements: Medication therapy review (MTR); personal medication record (PMR); medication-related action plan (MAP); intervention and/or referral; documentation and follow-up (Burns, 2008).

20.8 Online Patient Care Regulations

Online patient care is part of e-health regulations; development of international guidelines for online patient care by pharmacists is very important.

20.9. Facilitators for Effective Online Patient Care

20.9.1 Suitable and Effective Technologies and Tools

Suitable and effective technologies and tools are the keys to success in online patient care. Pharmacists should use and adapt the most effective technologies and tools for online patient care as follows:

Internet

The internet plays a very important and vital role in online patient care.

Computers and Laptops

Using Computers and laptops is very important and essential for effective online patient care.

Smartphones, Tablets, and Net Books

Using smartphones, tablets, and net books is very important and essential for effective online patient care.

Webinar and Video Conferencing Platforms

Webinar and video conferencing platforms are very important in online patient care. Microsoft Teams, Cisco WebEx Teams, Google Meet, and Zoom are very important for the success of online patient care. They include free versions that are great for light uses, short conference calls, and light file sharing.

WhatsApp

WhatsApp can be used in online patient care as an effective communication tool between pharmacists and patients.

Wearable Technologies

Wearable technologies are very important and essential for effective online patient care.

20.10 Barriers to Effective Online Patient Care

Lack of technology and infrastructure.

Lack of interest.

Knowledge of pharmacists.

Attitude of pharmacists.

Knowledge of public and patients.

Attitude of public and patients.

Financial issues.

Support.

Workforce issues.

Lack of time.

Lack of motivation.

Legal barriers.

20.11 Conclusion

This chapter has discussed the online patient care issues, facilitators, and barriers for implementing and providing online patient care services. Pharmacists have contributed effectively to patients' health; diseases/conditions management as well as prevention; treating outcomes; treating cost; patients' quality of life; and satisfaction towards the health care system and care. Pharmacists nowadays provide effective online services to patients and the public.

References

Afanasjeva, J., Burk, M., Cunningham, F., Fanikos, J., Gabay, M., Hayes, G., Masters, P.L., Rodriguez, R. and Sinnett, M.J., 2021. ASHP guidelines on medication-use evaluation. *American Journal of Health-System Pharmacy*, 78(2), pp. 168–175.

Al-Worafi, Y.M., 2020. Quality indicators for medications safety. In *Drug safety in developing countries* (pp. 229–242). Academic Press.

American Society of Health-System Pharmacists (ASHP), 1996. ASHP guidelines on medication-use evaluation. *American Journal of Health-System Pharmacy*, 53(16), pp. 1953–1955.

Burns, A., 2008. Medication therapy management in pharmacy practice: core elements of an MTM service model (version 2.0). *Journal of the American Pharmacists Association*, 48(3), pp. 341–353.

CMM in Primary Care Research Team., 2018. *The patient care process for delivering comprehensive medication management (CMM)*. American College of Clinical Pharmacy.

International Pharmaceutical Federation (FIP), 2020. *Medicines use review: A toolkit for pharmacists*. International Pharmaceutical Federation.

Kubacka, R.T., 1996. A primer on drug utilization review. *Journal of the American Pharmaceutical Association* (Washington, DC: 1996), 4, pp. 257–261.

World Health Organization (WHO), 2003. *Drug and therapeutics committees: A practical guide* (No. WHO/EDM/PAR/2004.1). World Health Organization.

21

Continuing Professional Development (CPD) and Lifelong Learning

21.1 Background

The concept of continuing professional development (CPD) can be defined as "the responsibility of individual pharmacists for systematic maintenance, development and broadening of knowledge, skills and attitudes, to ensure continuing competence as a professional, throughout their careers" (FIP, 2002). The Accreditation Council for Pharmacy Education (ACPE) defines CPD as "the lifelong process of active participation in learning activities that assists individuals in developing and maintaining continuing competence, enhancing their professional practice, and supporting achievement of their career goals" (ACPE, 2015). CPD stands for continuing professional development. It refers to the process of tracking and documenting the skills, knowledge, and experience that you gain both formally and informally as you work, beyond any initial training. It's a record of what you experience, learn, and then apply. The term is generally used to mean a physical folder or portfolio or e-portfolio documenting your development as a professional. Some organizations use it to mean a training or development plan, which I would argue is not strictly accurate. The purpose of CPD in health care is to help improve the safety and quality of care provided for patients and the public. As a health care professional you are responsible for identifying your CPD needs, planning how those needs should be addressed, and undertaking CPD that will support your professional development and practice. Your CPD activities should aim to maintain and improve the standards of your own practice and also those of any teams in which you work.

21.2 History

The history of continuing medical education (CME) goes back to 1935; however, only in the 1960s did CME start to be discussed as a coherent body of literature (Filipe et al., 2014; Grant, 2012).

DOI: 10.1201/9781003230458-23

21.3 Continuing Professional Development (CPD) and Lifelong Learning

Life-long learning can be defined as all learning activity undertaken throughout life, with the aim of improving knowledge, skills, and competence, within a personal, civic, social, and/or employment-related perspective (Alsop, 2013). Pharmacists by the nature of their education and practice during their B-Pharm or PharmD or others were graduated with the ability to be lifelong learners, which is a very important competency for them. Self-learning or self-directed learning was defined by Knowles as a process in which a learner takes the initiative, diagnoses their learning needs, creates learning goals, identifies resources for learning, applies appropriate learning strategies, and evaluates their learning outcomes (Knowles, 1975). The Accreditation Council for Pharmacy Education (ACPE) defines CPD as "the lifelong process of active participation in learning activities that assists individuals in developing and maintaining continuing competence, enhancing their professional practice, and supporting achievement of their career goals" (ACPE, 2015). The history of self-learning or self-directed learning goes back to the 1800s, In 1840 in the United States, Craik documented and celebrated the self-education efforts of several people showing early scholarly efforts to understand self-directed learning. In 1895 in Great Britain, Smiles published a book entitled *Self-Help*, that applauded the value of personal development (SDL Timeline, 2021). Pharmacy schools/departments should provide an environment and culture that promotes self-directed lifelong learning among pharmacy students to participate in self-learning/self-directed learning activities. Pharmacy schools/departments should ensure that the pharmacy curriculum includes self-learning/self-directed learning experiences and time for independent study to allow pharmacy students to develop and improve the required skills of lifelong learning. Preparing pharmacy students to be lifelong learners is very important and should be included in all pharmacy programs' competencies and learning outcomes.

21.4 Rationality of Continuing Professional Development (CPD) and Lifelong Learning

Pharmacy practice has changed during the last decades around the world and continues changing as a result of advances in medicine, health care and patient care practices, new technologies, systems, and process improvements, as well as changes in professional roles and responsibilities, which require pharmacists and future pharmacists to be ready for any change, and to have the required knowledge and skills to provide the most effective pharmacist care/patient care services. Nowadays all/the majority of pharmacy licensing bodies worldwide ask pharmacists and pharmacy professionals for evidence of continuing professional development (CPD) or continuing medical education (CME) to renew their license. CPD has the following benefits for pharmacists, organizations, and patients (DHA, 2014):

Benefits to the Patient

Patients receive safe, high-quality, and evidence-based service.

Benefits to the Professional

Improves confidence in delivery of professional service. Promotes and maintains competence to practice. Improves satisfaction with work role. Promotes lifelong learning. Provides structure and support for the health professional and for his or her valued goals. Enhances career opportunities.

Benefits to the Organization

Contributes to meeting the increasing demand for accountability, flexibility, and a skilled and competent workforce. Improves inter-professional working. Meets organizational objectives. Improves staff motivation and morale. Contributes to quality assurance.

Principles of Continuing Professional Development (CPD)

The principles of continuing professional development (CPD) can be summarized as follows (Rouse, 2004):

- CPD is a systematic, ongoing cyclical process of self-directed learning.
- It includes everything that practitioners learn that enables them to be more effective as professionals, that is, both traditional continuing education and other forms of professional development.
- CPD includes the entire scope of the practitioner's practice and it may include activities both within and outside the usual work setting.
- CPD is a partnership between the practitioner and his or her organization, meeting the needs of both.
- Practitioners are responsible for their own professional development. Organizations have a responsibility to help practitioners meet development needs that relate to performance in their current jobs.

Steps of Continuing Professional Development (CPD)

The steps of continuing professional development (CPD) can be summarized as follows (FIP, 2002):

1. **Self-Appraisal**—identification of continuing professional development needs may be accomplished by one or more of the following:
   - Personal assessment
   - Performance review by a manager
   - Audit exercise undertaken with others
   - Requirement of professional or health authority
2. **Personal Plan**—identify resources and actions required to meet personal CPD needs.
3. **Action**—participate in CPD (including presentations, tutoring, formal and informal meetings, workshops, short courses, teaching, talking with colleagues and experts, mentoring, formal education programs and self-study, among other methods).
4. **Documentation**—keep records of all CPD activities completed and provide that documentation when required.
5. **Evaluation**—evaluate personal benefit and benefit to patients from participation in any significant CPD activity. The following questions should be asked and answered:
   - Were the addressed needs met?
   - How has practice improved?
   - How have patients benefited?
   - Did learning break down? If so, why?

21.5 Barriers to Continuing Professional Development (CPD) and Online CPD

Lack of technology and infrastructure.

Lack of interest.

Knowledge of pharmacists.

Attitude of pharmacists.

Financial issues.

Support.
Lack of time.
Lack of motivation.
Cost.

21.6 Facilitators for Online Continuing Professional Development (CPD)

21.6.1 Internet

The internet plays a very important and vital role in online continuing professional development (CPD).

21.6.2 Computers and Laptops

Using computers and laptops is very important and essential for effective CPD.

21.6.3 Smartphones, Tablets, and Net Books

Using smartphones, tablets, and net books very important and essential for effective online CPD.

21.6.4 Webinar and Video Conferencing Platforms

Webinar and video conferencing platforms are very important in online CPD. Microsoft Teams, Cisco WebEx Teams, Google Meet, and Zoom are very important for the success of online patient care. They include free versions that are great for light uses, short conference calls, and light file sharing.

21.6.5 WhatsApp

WhatsApp can be used in online CPD as an effective communication tool.

21.6.6 Wearable Technologies

Wearable technologies are very important and essential for effective online CPD.

21.7 Online Continuing Professional Development (CPD)

Nowadays many local and international universities, organizations, publishers such as Elsevier, ASHP and others provide many certified courses with paid and free courses. The duration of certified courses can be days, weeks, or months depending on the type of certified course and program. Pharmacists can attend many CPD courses, workshops, and conferences.

21.7.1 Anticoagulation Certificate

Organized by the American Society of Health-System Pharmacists (ASHP): 32 continuous education (CE) hours designed for pharmacists to expand their knowledge and skills in anticoagulation therapy management across special populations and within acute care, ambulatory, and perioperative settings (ASHP, 2021).

21.7.2 Pharmacogenomics Certificate

Organized by the American Society of Health-System Pharmacists (ASHP): 20 continuous education (CE) hours designed for participants to increase the knowledge and skills necessary to use pharmacogenomics to improve medication use in a variety of patient care settings (ASHP, 2021).

21.7.3 Pain Management Certificate

Organized by the American Society of Health-System Pharmacists (ASHP): 21.5 continuous education (CE) hours designed for pharmacists to develop the knowledge and skills necessary to provide optimal pain management in patients suffering from chronic pain (ASHP, 2021).

21.7.4 Diabetes Management Certificate

Organized by the American Society of Health-System Pharmacists (ASHP): 33 continuous education (CE) hours designed to increase the knowledge and skills associated with the diagnosis, management, and pharmacological treatment of diabetes to optimize patient outcomes in ambulatory and inpatient care settings (ASHP, 2021).

21.7.5 Medication Safety Certificate

Organized by the American Society of Health-System Pharmacists (ASHP): 40 continuous education (CE) hours designed to enhance the skills and experience for pharmacy professionals, physicians, and nurses who lead medication safety improvements (ASHP, 2021).

21.7.6 Advanced Certificates and Board Certificates for Pharmacists

There are many advanced certificates and board certificates for pharmacists organized by many national and international organizations. Certain courses were equal to/accredited as professional master's degrees in many countries such as "Board Certified Pharmacotherapy Specialist (BCPS)."

The history of board certificates goes back to the early 1970s by the American Pharmacists Association (APhA).

Advanced certificates and board certificates are contributing effectively to the improvement of pharmacist care/patient care services and improving pharmacists' knowledge and skills towards advanced services. Examples of Advanced Certificates and Board Certificates are:

21.7.7 The Pharmacist Independent Prescribing Practice Certificate

The Pharmacist Independent Prescribing Practice Certificate is provided by many universities in the United Kingdom (UK). This program/certificate aims to improve the registered pharmacist's knowledge and skills towards prescribing and prepare them to be good pharmacist prescribers.

21.7.8 Board Pharmacy Specialties (BPS)

The BPS Board Certified Nuclear Pharmacist specializes in the procurement, preparation, compounding, dispensing, and distribution of radiopharmaceuticals, as well as the regulatory aspects governing these processes. In addition, the nuclear pharmacist serves as the medication expert within the health care team regarding clinical aspects of radiopharmaceuticals and nonradioactive drugs used in patient care (BPS, 2021).

21.7.9 Board Certified Medication Therapy Management Specialist (BCMTMS)

Board Certified Medication Therapy Management Specialists (BCMTMS) are organized by the National Board of Medication Therapy Management (NBMTM), designed to improve the pharmacist's knowledge and skills towards medication therapy management and prepare them to provide effective medication therapy management to patients.

21.7.10 Oncology Pharmacy Board Certificate

Oncology Pharmacy Board Certificates provide evidence-based, patient-centered medication therapy management and direct patient care for individuals with cancer, including treatment assessment and monitoring for potential adverse drug reactions and interactions (BPS, 2021).

21.7.11 Solid Organ Transplantation Pharmacy Certificate

Solid Organ Transplantation Pharmacy Certificates provide evidence-based, patient-centered medication therapy management and care for patients throughout all phases of solid organ transplantation at all ages and in various health care settings (BPS, 2021).

21.7.12 Ambulatory Care Pharmacy Certificate

Ambulatory Care Pharmacy Certificates address the provision of integrated, accessible health care services for ambulatory patients in a wide variety of settings, including community pharmacies, clinics, and physician offices (BPS, 2021).

21.7.13 Compounded Sterile Preparations Pharmacy Certificate

Compounded Sterile Preparations Pharmacy Certificates ensure that sterile preparations meet the clinical needs of patients, satisfying quality, safety, and environmental control requirements in all phases of preparation, storage, transportation, and administration in compliance with established standards, regulations, and professional best practices (BPS, 2021).

21.8 Free Online Continuing Professional Development (CPD)

The cost of continuing professional development (CPD) is very important for pharmacists and one of the major barriers for attending CPDs among pharmacists. On the other hand, free CPDs are one important key for motivating pharmacists and pharmacy professionals to attend CPDs. Many organizations provide free online CPD for pharmacists and health care professionals with continuing medical education (CME) and without CME, which are available on the following websites by well-known organizations such as the American Society of Health-System Pharmacists (ASHP):

elearning.ashp.org/catalog/free

Medscape Education | Pharmacists-=

Online Health & Medicine Courses | Harvard University

https://waterfalls.ae/

21.9 Conclusion

This chapter has discussed the history of continuing professional development (CPD), its importance and the facilitators and barriers for attending the continuing professional development (CPD). Pharmacy practice has changed during the last decades around the world and continues changing as a result of advances in the medicine, health care and patient care practices, new technologies, systems, and process improvements, as well as changes in professional roles and responsibilities, which require pharmacists and future pharmacists to be ready for any change, and to have the required knowledge and skills to provide the most effective pharmacist care/patient care services. Nowadays all/the majority of pharmacy licensing bodies worldwide ask pharmacists and pharmacy professionals for evidence of continuing professional development (CPD) or continuing medical education (CME) to renew their license.

References

Accreditation Council for Pharmacy Education (ACPE), 2015. *Accreditation standards and key elements for the professional program in pharmacy leading to the doctor of pharmacy degree* (Standards 2016). Available at: https://www.acpe-accredit.org/pdf/Standards2016FINAL.pdf

ASHP, 2021. Available at: https://elearning.ashp.org/

Alsop, A., 2013. *Continuing professional development in health and social care: Strategies for lifelong learning.* John Wiley & Sons.

BPS, 2021. Available at: https://www.bpsweb.org/

Dubai Health Authority (DHA), 2014. *Continuing Professional Development (CPD) guidelines.* Available at: https://www.dha.gov.ae/Documents/About%20DHA/Continuing%20Professional%20Development%20Guidline.pdf

Filipe, H.P., Silva, E.D., Stulting, A.A. and Golnik, K.C., 2014. Continuing professional development: Best practices. *Middle East African Journal of Ophthalmology,* 21(2), p. 134.

Grant J., 2012. *The good CPD guide: A practical guide to managed continuing professional development in medicine.* 2nd ed. New York Radcliffe Publishing.

International Pharmaceutical Federation (FIP), 2002. *FIP statement of professional standards: continuing professional development.* Available at: https://www.fip.org/file/1544

Knowles, M.S., 1975. *Self-directed learning.* Prentice Hall Regents.

Rouse, M.J., 2004. Continuing professional development in pharmacy. *Journal of Pharmacy Technology,* 20(5), pp. 303–306.

SDL Timeline., 2021. *Historical timeline for self-directed learning in adult education.* Available at: http://sdlearning.pbworks.com/w/page/1899125/SDL%20Timeline

22

Pharmacists' Prescribing

22.1 Terminologies

22.1.1 Prescribing

"To give directions, either orally or in writing, for the preparation and administration of a remedy to be used in the treatment of any disease" (Stedman, 1920).

22.1.2 Diagnosing

The determination of the nature of a disease, injury, or congenital defect (Stedman, 1920).

22.1.3 Nonmedical Prescribing

Prescribing by specially trained nurses, optometrists, pharmacists, physiotherapists, podiatrists, radiographers, and dietitians working within their clinical competence as either independent and/or supplementary prescribers (RPS, 2016).

22.1.4 Independent Prescribing

Independent prescribing is prescribing by a practitioner who is responsible and accountable for the assessment of patients with undiagnosed or diagnosed conditions and for decisions about the clinical management required, including prescribing. In practice, there are TWO distinct forms of nonmedical independent prescriber:

i) At time of publication an independent prescriber may be a specially trained nurse, pharmacist, optometrist, physiotherapist, therapeutic radiographer, or podiatrist who can prescribe licensed medicines within their clinical competence. Nurse and pharmacist independent

DOI: 10.1201/9781003230458-24

prescribers can also prescribe unlicensed medicines and controlled drugs.

ii) A community practitioner nurse prescriber (CPNP), for example, a district nurse, health visitor, or school nurse, can independently prescribe from a limited formulary called the Nurse Prescribers' Formulary for Community Practitioners, which can be found in the British National Formulary (BNF) (RPS, 2016).

22.1.5 Supplementary Prescribing

Supplementary prescribing is a voluntary partnership between a doctor or dentist and a supplementary prescriber to prescribe within an agreed patient-specific clinical management plan (CMP) with the patient's agreement. Nurses, optometrists, pharmacists, physiotherapists, podiatrists, radiographers, and dietitians may become supplementary prescribers and once qualified may prescribe any medicine within their clinical competence, according to the CMP (RPS, 2016).

22.1.6 Dependent Prescribing/Collaborative Prescribing

Dependent prescribing/collaborative prescribing is when a pharmacist strikes an agreement with a physician. This agreement outlines which patients the pharmacist can prescribe for (usually it would be the patients under the care of that physician) and what the pharmacist can prescribe (e.g., metformin for patients with newly diagnosed type 2 diabetes).

22.2 History

Traditionally, in the United Kingdom and many other parts of Europe and the world, groups of tradespeople called chemists, druggists, and apothecaries have performed quite similar tasks. Over time, apothecaries split from the chemists and druggists to become general practitioners, while the other two groups combined to become pharmacists. Therefore, moving into prescribing may be just a matter of returning to the beginnings of our trade (Malleck, 2004). Pharmacists prescribing in modern history goes back to 2003 in the UK followed by other countries (Force, 2010). Historically, literature reported that patients have relied on physicians and dentists to prescribe medications, order laboratory tests, and conduct or supervise procedures consistent with the patients' diagnoses. More recently, prescribing privileges have been extended to other health care professionals, such as nurse practitioners, nurses with an expanded scope of practice, clinical nurse specialists, registered midwives, dieticians, podiatrists, optometrists, and, in many countries, pharmacists (Force, 2010). However, pharmacists in many developing countries prescribe illegally due to several factors, as reported in Yemen: most Yemenis do not go to physicians because they cannot afford the treatment in private hospitals or clinics, and generally there is no medical insurance. Unfortunately, the government hospitals and medical centers are the worst in the country. Also, care in these facilities is reserved for a patient who has a relative in the hospital or is affiliated with influential people. Further, patients are required to buy everything from the papers used to write prescriptions to the medicines. People are also unaware of the dangers of buying prescription medicines without a valid prescription. The Ministry of Public Health and Population and the Supreme Board of Drugs and Medical Appliances fail to regulate, control, and monitor the prescriptions. No policy exists requiring dispensers to be qualified and registered as pharmacists. Anyone with money in Yemen can open a pharmacy by renting a license from a pharmacist and hiring nonqualified pharmacists to work in the pharmacy (Fathelrahman et al., 2016; Al-Worafi, 2016, 2014). Nesbitt (2012) reported that the history of telemedicine goes back to the late 1950s when one of the earliest and most famous uses of hospital-based telemedicine was in the late 1950s and early 1960s where a closed-circuit television link was established between the Nebraska Psychiatric Institute and Norfolk State Hospital for psychiatric consultations (Nesbitt, 2012). During the 2000s and forwards, after the evolution of the internet, telemedicine was implemented in many countries around the world, especially after the discovery of mobile technologies and mobile applications.

22.3 Rationality of Pharmacists' Prescribing/ Online Prescribing

During 2020 and 2021 the majority of outpatient clinics were closed in many countries around the world, which lead to increasing the demand for e-health with the help of technology. New technologies played a very important role to facilitate e-health. Pharmacists during the lockdown and forwards used the new technologies to provide effective online pharmacist care and patient care services. Pharmacy practices in developed and many developing countries have witnessed many changes during the last decades towards improving pharmacist care/patient care services. Revolutionary changes in pharmacy practice a few decades ago, when the pharmacy profession witnessed great practice changes and moved away from its original focus on medicine supply and dispensing towards a focus on patient care, especially after the introduction of clinical pharmacy concepts in the late 1960s, followed by the philosophy of pharmaceutical care in the early 1990s. However, pharmacists nowadays play/should play an important role in patient care. They contribute effectively to health, diseases/conditions management, as well as prevention; treating outcomes; treating cost; patients' quality of life; satisfaction towards the health care system and care. Independent-prescribing models are part of a pharmacist's full scope of practice. Pharmacists are valuable members of the health care team and among the most accessible health care providers. Pharmacists and pharmacy organizations need to have the courage to pursue their essential place as part of the health care team and should not be asking for permission to care for their patients (Tsuyuki et al., 2018; Tsuyuki and Watson., 2020). Pharmacists' prescribing contributed/will contribute to the patients' clinical, economic, and humanistic treatment outcomes, decrease the pressure on hospitals, improve patient care, improve access to medication, and optimize medication management and better resource utilization (Emmerton et al., 2005; Hoti et al., 2011; Tonna et al., 2007). Online prescribing/telemedicine play an important role in providing patient care services to remote areas or areas with a shortage of health care providers such as in many developing countries.

22.4 Quality of Pharmacists' Prescribing

Good prescribing has a good impact on patients' health as well as the health care system. Good quality prescribing, prescribing effective and safe medications, and good prescribing will help in achieving treatment outcomes; decreasing the admission rate to hospitals; decreasing morbidity and mortality; decreasing the cost of therapy; improving the quality of life and improving patients' satisfaction towards health care. Prescribing quality indicators are very important and necessary in order to evaluate the prescribing practice, identify the problems as well as challenges, and develop and implement action plans to overcome the identified challenges and improve the practice. Prescribing quality indicators comprise five sections as follows (Al-Worafi, 2020):

Section 1. Communication Skills Indicators

Did the prescriber:

Welcome the patient?

Introduce him/herself?

Explain the aim of conversation/gathering information?

Speak clearly?

Make eye contact?

Check whether you understood what you had been told?

Ask if the patient needed additional information or had questions?

Section 2. Gathering Information Indicators

Did the prescriber gather/collect the following information?

Patients' related information: gender, age, weight, height.

Chief complaint.

History of present illness.

Medical history.

Adherence.

Medications history (prescribed, OTC/self-medications, herbal).

Allergies to medications, herbals, foods, others.

Family history.

Surgical history.

Social history (marital status, number of children, educational level, occupation, smoking, alcohol, "lifestyle" such as exercise, eating habits).

Review of Systems: from head to toe which include: general appearance and health status; skin (integumentary); vital signs (VS): blood pressure, heart rate, temperature, respiration; HEENT: head, ears, eyes, nose, throat; lungs/thorax (pulmonary); cardiovascular; abdomen; genit/rect (genitalia/rectal); MS/Ext (musculoskeletal and extremities); NEURO: neurological exam and PSYCH: mental status exam.

Section 3. Diagnosis Indicators

Did the prescriber request/recommend?

The needed laboratory tests.

Other diagnosis requirements.

Section 4. Management Indicators

Did the prescriber develop/implement the following for each disease/condition?

Goals of therapy and desired outcomes for all diseases/conditions.

Nonpharmacological therapies (individualize lifestyle changes such as weight control, healthy dietary therapy, increased physical activity, modifying the modifiable risk factors, etc.) depending on the disease and the patient's situation.

Pharmacological therapies (appropriate and rational based on the guidelines) recommendations with doses, dosage form and route of administration, strength, frequency, duration; time of taking medications and instructions.

Section 5. Monitoring Parameters

The efficacy of medications (is the prescribed medication effective; is the desired outcome achieved). This can be done by using the laboratory results, checking the symptoms' improvement, patients' report, and other criteria.

The safety of medications (is the prescribed medication safe). This can be done by patients' reports about side/adverse effects/reactions, evaluating the effects on the patient's different systems such as renal, liver, and so on, requesting laboratory tests, requesting drug therapy monitoring (DTM), and others.

Adherence towards the management plan.

Therapy success and complications: Is the treatment's desired outcome achieved?

Patient education and counseling related to the adherence towards the management plan (Nonpharmacological, pharmacological therapies, and monitoring parameters), self-management, potential adverse drug effects and reactions, possible interactions, cautions and precautions, contraindications and warnings, proper storage and disposal of medications.

22.5 Competencies

The Royal Pharmaceutical Society (RPS, 2016) developed a competency framework for all prescribers as follows: (1) assess the patient; (2) consider the options; (3) reach a shared decision; (4) prescribe; (5) provide information; (6) monitor and review; (7) prescribe safely; (8) prescribe professionally; (9) improve prescribing practice; and (10) prescribe as part of a team (Royal Pharmaceutical Society (RPS, 2016). The National Prescribing Service (NPS, 2012) in Australia developed competencies required to prescribe medicines as follows:

Competency Area 1

Understands the person and their clinical needs.

Competency Area 2

Understands the treatment options and how they support the person's clinical needs.

Competency Area 3

Works in partnership with the person to develop and implement a treatment plan.

Competency Area 4

Communicates the treatment plan clearly to other health professionals.

Competency Area 5

Monitors and reviews the person's response to treatment.

Horizontal Competency Area H1

Practices professionally.

Horizontal Competency Area H2

Communicates and collaborates effectively with the person and other health professionals.

The horizontal competencies (H1 and H2) are competencies that health professionals integrate with the other competency areas during the prescribing cycle (NPS, 2012).

22.6 Facilitators for Effective Online Prescribing

Suitable and effective technologies and tools
Internet
Computers and laptops
Smartphones, tablets, and net books
Webinar and video conferencing platforms
WhatsApp
Wearable technologies
Wearable technologies are very important and essential for effective online patient care.

22.7 Barriers to Effective Online Prescribing

Lack of technology and infrastructure.
Lack of interest.

Knowledge of pharmacists.

Attitude of pharmacists.

Knowledge of the public and patients.

Attitude of the public and patients.

Financial issues.

Support.

Workforce issues.

Lack of time.

Lack of motivation.

Legal barriers.

22.8 Pharmacists' Prescribing/Online Prescribing Regulations

Pharmacists' prescribing/online pharmacists' prescribing, or telemedicine regulations are very important to ensure the efficacy and safety of prescribing. Many countries have developed their regulations such as the UK, the US, Australia, and others, while many countries, especially developing countries, don't have regulations till now. However, adapting the international regulations can help these countries to organize online prescribing and telemedicine. Minimum requirements for pharmacists' prescribing should include: the pharmacist should be licensed and his/her license in good status with the licensing bodies; training about prescribing and getting a certificate about it such as the Pharmacist Independent Prescribing Practice Certificate which is provided by many universities in the United Kingdom (UK). This program/certificate aims to improve the registered pharmacist's knowledge and skills towards prescribing and prepare them to be good pharmacist prescribers, available of the minimum facilities.

22.9 Conclusion

This chapter has discussed the online pharmacists' prescribing. This chapter includes the terminologies related to the pharmacists' prescribing; the history of online prescribing; the rationality of pharmacists' prescribing/ online prescribing; the quality of pharmacists' prescribing and the required competencies; barriers to and facilitators for effective online prescribing and pharmacists' prescribing regulations. Pharmacists' prescribing has contributed/will contribute to patients' clinical, economic, and humanistic treatment outcomes, decrease the pressure on hospitals, improve patient care, improve access to medication, optimize medication management and better resource utilization. Online prescribing/telemedicine plays an important role in providing patient care services to remote areas or areas with a shortage of health care providers such as in many developing countries.

References

Al-Worafi, Y.M., 2014. Pharmacy practice and its challenges in Yemen. *The Australasian Medical Journal*, 7(1), p. 17.

Al-Worafi, Y.M.A., 2016. Pharmacy practice in Yemen. In *Pharmacy practice in developing countries* (pp. 267–287). Academic Press.

Al-Worafi, Y.M., 2020. Quality indicators for medications safety. In *Drug safety in developing countries* (pp. 229–242). Academic Press.

Emmerton, L., Marriott, J., Bessell, T., Nissen, L. and Dean, L., 2005. Pharmacists and prescribing rights: Review of international developments. *Journal of Pharmacy and Pharmaceutical Sciences*, 8(2), pp. 217–225.

Fathelrahman, A., Ibrahim, M. and Wertheimer, A., 2016. *Pharmacy practice in developing countries: Achievements and challenges*. Academic Press.

Force, P.P.T., 2010. Prescribing by pharmacists: information paper (2009). *The Canadian Journal of Hospital Pharmacy*, 63(3), p. 267.

Hoti, K., Hughes, J. and Sunderland, B., 2011. An expanded prescribing role for pharmacists-an Australian perspective. *The Australasian Medical Journal*, 4(4), p. 236.

Malleck, D.J., 2004. Professionalism and the boundaries of control: Pharmacists, physicians and dangerous substances in Canada, 1840–1908. *Medical History*, 48(2), pp. 175–198.

Nesbitt, T.S., 2012. The evolution of telehealth: where have we been and where are we going. In Board on Health Care Services and Institute of Medicine, eds. *The role of telehealth in an evolving health care environment: Workshop summary*. National Academies Press (US).

Tsuyuki, R.T. and Watson, K.E., 2020. *Why pharmacist prescribing needs to be independent*. Available at: https://www.ncbi.nlm.nih.gov/pmc/articles/PMC7079324/

Tsuyuki, R.T., Houle, S.K. and Okada, H., 2018. *Time to give up on expanded scope of practice*. Available at: https://journals.sagepub.com/doi/full/10.1177/1715163518793844

NPS: Better choices, Better health., 2012. *Competencies required to prescribe medicines: Putting quality use of medicines into practice*. National Prescribing Service Limited. Available at: 682949fec05647bc-2c0de122631e-Prescribing_Competencies_Framework.pdf (nps.org.au)

Royal Pharmaceutical Society (RPS), 2016. *A competency framework for all prescribers.* Available at: www.rpharms.com/Portals/0/RPS%20document%20library/ Open%20access/Professional%20standards/Prescribing%20competency%20 framework/prescribing-competency-framework.pdf

Stedman, T., 1920. *Stedman's medical dictionary.* Dalcassian Publishing Company.

Tonna, A.P., Stewart, D., West, B. and McCaig, D., 2007. Pharmacist prescribing in the UK: A literature review of current practice and research. *Journal of Clinical Pharmacy and Therapeutics,* 32(6), pp. 545–556.

23

Advantages, Disadvantages, and Quality Issues

23.1 Advantages of Online Pharmacy Practice

Distance/remote and online pharmacist care has many benefits for patients and the public, especially those in rural areas or those who can't access or reach the pharmacies and pharmacists. Using technology to deliver pharmaceutical care and patient care services has several advantages, including cost savings, convenience, and the ability to provide care to people with mobility limitations, or those in rural areas who don't have access to a local doctor or clinic.

23.1.1 Access to Health Care Services

Online pharmacy services have helped/can help to treat patients at their homes like what happened during the lockdown and after COVID-19 when the majority of outpatient clinics were closed in many countries around the world, which lead to an increased demand for e-health, with the help of technology and new technologies playing a very important role to facilitate e-health. Pharmacists during the lockdown and forwards used the new technologies to provide effective online pharmacist care and patient care services. People can access the online services more easily than going to the pharmacies.

23.1.2 Safety

Online pharmacy practice is safer than face-to-face practice.

23.1.3 Cost Savings

Patients and the public can save money directly and indirectly with online pharmacist care and patient care services.

DOI: 10.1201/9781003230458-25

23.1.4 Convenience

Online pharmacist care and patient care services can be more convenient for patients.

23.1.5 Improve Technology Knowledge

Online pharmacy education is a good opportunity for pharmacists and patients/the public to improve their knowledge towards technologies and new technologies.

23.1.6 Improve Technology Skills

Online pharmacy education is a good opportunity for pharmacists and patients/the public to improve their skills towards technologies and new technologies.

23.1.7 Environment

Online pharmacy practice has benefits for the environment.

23.1.8 Benefits to Pharmacists

Online pharmacy practice has benefits to pharmacists as follows:

Cost saving: Online pharmacy practice requires less infrastructure than traditional education.

Earn money: Online pharmacy practice can be a good income source for pharmacists.

Safety: Online pharmacy practice is safer for pharmacists than face-to-face practice.

Satisfaction: Online pharmacy practice can improve satisfaction towards pharmacist care services and helping patients and the public.

Practice: Online pharmacists can practice new technologies while providing online patient care and services.

23.2 Disadvantages and Problems of Online Pharmacy Practice

Online pharmacy practice has many disadvantages and problems as follows:

Lack of Infrastructure and Resources

Many pharmacies and pharmacists around the world, especially in low-income and middle-income countries, faced problems such as the following:

Internet

The internet plays a very important and vital role in online pharmacy practice. Many pharmacists and patients/the public especially in low-income and middle-income countries face the problem of lack of access to the internet.

Computers and Laptops

Using computers and laptops is very important and essential for online pharmacy practice. It provides flexible and effective access to online services. Many pharmacists and patients/the public especially in low-income and middle-income countries lack access.

Smartphones, Tablets, and Net Books

Using smartphones, tablets, and net books is very important and essential for online pharmacy practice. It provides flexible and effective access to online services. Many pharmacists and patients/the public especially in low-income and middle-income countries lack access.

Lack of/Poor Communication Skills

Many pharmacists as well as patients have this problem.

Lack of/Poor Technology Use Skills

Many pharmacists as well as patients have this problem.

Workload

Many pharmacists' workload has been increased due to lack of staff.

Lack of/Insufficient Training

Many pharmacists face this problem.

Lack of/Insufficient Technical Support

Many pharmacists face this problem.

23.3 Quality of Online Dispensing

Dispensing medications practice nowadays has changed from the product orientation towards a patient care orientation (Hepler, 2004; Strand et al., 2012; Joint, 2011) In order to provide good online dispensing, pharmacists should have/do the following (Al-Worafi, 2020a):

I. Prescribed Medications

Good communication skills

Check the prescriptions/orders quality and appropriateness

Appropriate dispensing

Appropriate labeling

Appropriate packaging

Appropriate patient education and counseling

II. Nonprescription Medications (OTC)/Self-Medication Practice

Good communication skills

Gather the patient's related information

Patient assessment and diagnosis

Appropriate management: goals of therapy and desired outcomes for all diseases/conditions.

Nonpharmacological therapies (individualize " lifestyle changes" such as weight control, healthy dietary therapy, increased physical activity, modifying the modifiable risk factors, and so on depending on the disease and patient situation). Pharmacological therapies (appropriate and rational based on the guidelines) recommendations with doses, strength, dosage form, and route of administration, frequency, duration; time of taking medications and instructions.

Appropriate dispensing

Appropriate labeling

Appropriate packaging

Appropriate monitoring parameters: The efficacy of medications (is the dispensed medication effective; is the desired outcome achieved). This can be done by using/recommending laboratory results, checking the symptoms improvement, or by patients' report and other criteria. The safety of medications (is the dispensed medications safe). This can be done by patients' reports about side/adverse effects/reactions, evaluating the effects on patients' different systems such as renal, liver, and so on, requesting/recommending laboratory tests, requesting drug therapy monitoring (DTM), and others. Adherence towards the management plan.

Therapy success and complications: Is the treatment desired outcome achieved?

Appropriate patient education and counseling.

Appropriate referral decision: refer patients to physicians, clinics, or hospitals based on the guidelines.

23.4 Quality of Online Prescribing

Good prescribing has a good impact on patients' health as well as the health care system. In order to provide good online dispensing, pharmacists should have/do the following (Al-Worafi, 2020a):

Good communication skills.

Gather the patient's related information.

Patient assessment and diagnosis.

Appropriate management: goals of therapy and desired outcomes for all diseases/conditions.

Nonpharmacological therapies (individualize "lifestyle changes" such as weight control, healthy dietary therapy, increased physical activity, modifying the modifiable risk factors, and so on depending on the disease and patient situation). Pharmacological therapies (appropriate and rational based on the guidelines) recommendations with doses, strength, dosage form and route of administration, frequency, duration, time of taking medications, and instructions.

Appropriate monitoring parameters: The efficacy of medications (is the dispensed medication effective; is the desired outcome achieved). This can be done by using/recommending laboratory results, checking the symptoms improvement, patients' report, and other criteria. The safety of medications (is the dispensed medications safe). This can be done by patients' reports about side/adverse effects/reactions, evaluating the effects on patients' different systems such as renal, liver, and so on, requesting/recommending laboratory tests, requesting drug therapy monitoring (DTM), and others. Adherence towards the management plan.

Therapy success and complications: Is the treatment desired outcome achieved?

Appropriate patient education and counseling.

23.5 Quality of Online Medications Safety Practice

Ensuring the safety of medications during online practice is very important and highly recommended. In order to provide good online pharmacy practice pharmacists should have/do the following (Al-Worafi, 2020b):

Educate and counsel patients about adverse drug reactions (ADRs), how they can prevent/minimize them, how they can report them directly to the national pharmacovigilance centers or discuss them with the pharmacists, then the pharmacists report them directly to the national pharmacovigilance centers or publish them as a case report.

Treat the actual ADRs.

Educate and counsel patients about medication errors, how they can prevent/minimize them, how they can report them directly to the national pharmacovigilance centers or discuss them with the pharmacists, then the pharmacists report them directly to the national pharmacovigilance centers or publish them as a case report.

Treat the actual medication errors.

Dispense antibiotics and other prescribed medications with valid prescriptions only.

Educate and counsel patients/the public about the appropriate use of their self-medications.

Identify patients/the public at risk for medications misuse and abuse, educate and counsel them about the effects of medications misuse and abuse on their health.

Fight the substandard and counterfeit medications, medical devices, cosmetics, and vaccines and report them to the national pharmacovigilance centers.

Educate and counsel patients/the public about the appropriate way to store their medications.

Educate and counsel patients/the public about the appropriate way to dispose of expired and unused medications.

23.6 Conclusion

This chapter has discussed the advantages and disadvantages of online pharmacy practice and services. Furthermore, it describes the quality of online pharmacy practice services. Distance/remote and online pharmacist care has many benefits to patients and the public, especially those in rural areas or those who can't access or reach pharmacies and pharmacists. Using technology to deliver pharmaceutical care and patient care services has several advantages, including cost savings, convenience, and the ability to provide care to people with mobility limitations, or those in rural areas who don't have access to a local doctor or clinic.

References

Al-Worafi, Y.M., 2020a. Quality indicators for medications safety. In *Drug safety in developing countries* (pp. 229–242). Academic Press.

Al-Worafi, Y.M., 2020b. Drug safety in developing versus developed countries. In *Drug safety in developing countries* (pp. 613–615). Academic Press.

Hepler, C.D., 2004. Clinical pharmacy, pharmaceutical care, and the quality of drug therapy. *Pharmacotherapy: The Journal of Human Pharmacology and Drug Therapy*, 24(11), pp. 1491–1498.

Joint, F.I.P., 2011. *WHO guidelines on good pharmacy practice: standards for quality of pharmacy services*. WHO technical report series, 961.

Strand, L.M., Cipolle, R.J. and Morley, P.C., 2012. *Pharmaceutical care practice*. McGraw-Hill.

Section 3

Online Pharmacy Research

24

History and Importance

24.1 History of Medical Research

The history of clinical research includes examples such as the following (Imarc, 2020; Sims et al., 2021; Lombard et al., 2007; Bhatt, 2010; Collier, 2009):

605 BC Book of Daniel

In one of the earliest recorded examples of classical experimental design in history, Daniel sought to test the effects and benefits of a vegetarian diet. He ate only vegetables while the other subjects enjoyed the King's meat and wine. After ten days, they assessed who was healthier—in this case, Daniel.

1537 First Clinical Trial of a Novel Therapy

The first clinical trial of a novel therapy was conducted accidentally by the famous surgeon Ambroise Pare in 1537.

1747 Clinical Trial of the Modern Era

James Lind is considered the first physician to have conducted a controlled clinical trial of the modern era.

1799 Smallpox Vaccine

Benjamin Waterhouse introduces the smallpox vaccine to the United States and helps gain acceptance for the new procedure.

1800

Arrival of the placebo.

DOI: 10.1201/9781003230458-27

Vaccine Discovery

Louis Pasteur's experiments spearheaded the development of live attenuated cholera vaccine and inactivated anthrax vaccine in humans (1897 and 1904, respectively).

1928 Sir Alexander Fleming Discovers Penicillin

Alexander Fleming's cluttered and untidy lab yielded one of the most important discoveries in the history of medicine as penicillin was identified on mold growing on a stack of staphylococci cultures. Soon thereafter, penicillin was recognized as one of the most efficacious life-saving drugs in the world, forever changing the treatment of bacterial infections.

1921 Insulin Discovery

Insulin was discovered by Sir Frederick G Banting, Charles H Best, and JJR Macleod at the University of Toronto in 1921 and it was subsequently purified by James B Collip.

1940s

The First Double Blind Controlled Trial—Patulin for Common Cold. The idea of randomization was introduced in 1923. However, the first randomized controlled trial of streptomycin in pulmonary tuberculosis was carried out in 1946.

1947 Nuremburg Code

During the Nuremburg Trials, 23 members of the German Nazi Party were tried for crimes against humanity for the atrocious experiments conducted on unwilling prisoners of war. The resulting verdict contained a set of 10 ethical principles for human experimentation. This Code established the requirements for informed consent, absence of coercion, and properly formulated scientific experimentation.

1964, Declaration of Helsinki

The Declaration expands on the 10 principles necessary for ethical human experimentation set forth in the Nuremberg Code. Developed by the World Medical Association, the Declaration of Helsinki is generally regarded as the cornerstone document of human research ethics. Since its inception in June 1964, it has undergone six revisions, the most recent coming in 2008. The focus of the Declaration is on respect for the subject, the subject's right to self-determination, and their right to make informed decisions regarding research participation. It stands as a clear and powerful statement that the rights of the human subject shall never be compromised for the sake of science and promotes beneficence towards experiment participants.

24.2 History of Online Research

The history of online research includes examples such as the following (Hooley et al., 2012; Thach, 1995; Kiesler and Sproull, 1986):

1986

Sproull (1986) lists four characteristics of electronic mail that make it useful for communication, and specifically for survey research: (1) Speed—messages can be transmitted in seconds to any location in the world, depending on the scope of the network. (2) Asynchronous communication—messages can be sent, read, and replied to at the convenience of the user. (3) No intermediaries—email messages are generally only read by the receiver. (4) Ephemerality— email messages appear on screen and can easily be deleted with no trace of a hard copy. This lends an ephemeral quality to electronic mail that cannot be matched by traditional mail. However, users still have the option of saving email messages in electronic folders and printing them out in hard copy if so desired. These four characteristics relate directly to the success of email survey research.

1994: Foster conducts online asynchronous interview using email.

1994: First methodological discussion of online interviewing (Brotherson, 1994).

24.3 Importance of Online Research in Pharmacy Education and Practice

Research in general about pharmacy education and practice-related issues is very important and necessary in order to evaluate the practice, identify the problems as well as challenges, and develop and implement action plans to overcome the identified challenges and improve the practice. Without research, the practice will not improve. Online research has many advantages such as the following:

Low Cost

Qualitative and quantitative online research is less expensive than traditional research. Researchers can save money directly and indirectly with online research.

Saving Time

Researchers can save time in travel and other ways with online research.

Sample Size/Number of Participants

Researchers can send online surveys to thousands of people in different countries quickly.

Improved Access to Populations

Researchers can access different populations quickly with online research, or can send a link to a survey via email, WhatsApp, and social media to participants. Respondents then have a variety of ways to access the questionnaire including mobile devices, tablets, laptops, desktop computers, and others.

Flexibility

Online surveys provide more flexibility in the design.

Convenience

It is very easy and convenient for respondents to complete surveys online.

Anonymity

Online surveys provide respondent anonymity.

Data Accuracy

The data collected through online surveys is more accurate. This is attributed to the fact that the responses go directly to the online database rather than being manually entered by participants.

Data Analysis

Many survey platforms can analyze the data quickly and immediately after receiving the surveys from the participants.

24.4 Disadvantages of Online Research in Pharmacy Education and Practice

There are disadvantages of online research such as the following:

Sampling Techniques

While online research can reach thousands of participants, there is still the problem of the randomization of sample size.

Technical Problems

Technical problems can interfere with respondents filling out online surveys.

Number of Questions

There are limited questions allowed for many free online survey platforms. Researchers can therefore not include many important questions.

Number of Participants

There are a limited number of participants allowed for many free online survey platforms. The researchers cannot invite and include more participants, which affects the sample size.

Equity Access

Many researchers and participants do not have the facilities for conducting online research, especially in many low-income and middle-income countries.

Bias

There is risk for bias in many online surveys.

Follow-Up

Researchers cannot conduct follow-up studies for much online research, because they don't have the participants' contact information.

Facilitators are needed for effective online pharmacy research implementation.

24.5 Conclusion

This chapter has discussed the history of medical research in general as well as the history of online research. It describes the importance of pharmacy research and online research. Furthermore, it describes the disadvantages of online research.

References

Bhatt, A., 2010. Evolution of clinical research: A history before and beyond James Lind. *Perspectives in Clinical Research*, 1(1), p. 6.

Brotherson, M.J., 1994. Interactive focus group interviewing: A qualitative research method in early intervention. *Topics in Early Childhood Special Education*, 14(1), pp. 101–118.

Collier, R., 2009. *Legumes, lemons and streptomycin: A short history of the clinical trial.* Available at: https://www.cmaj.ca/content/180/1/23

Hooley, T., Wellens, J. and Marriott, J., 2012. *What is online research?: Using the internet for social science research.* A&C Black.

Imarc., 2020. *History of clinical research.* Available at: www.imarcresearch.com/clinical-research-timeline-history

Kiesler, S. and Sproull, L.S., 1986. Response effects in the electronic survey. *Public Opinion Quarterly*, 50(3), pp. 402–413.

Lombard, M., Pastoret, P.P. and Moulin, A.M., 2007. A brief history of vaccines and vaccination. *Revue Scientifique et Technique-Office International des Epizooties*, 26(1), pp. 29–48.

Sims, E.K., Carr, A.L., Oram, R.A., DiMeglio, L.A. and Evans-Molina, C., 2021. 100 years of insulin: Celebrating the past, present and future of diabetes therapy. *Nature Medicine*, 27(7), pp. 1154–1164.

Sproull, L.S., 1986. Using electronic mail for data collection in organizational research. *Academy of Management Journal*, 29(1), pp. 159–169.

Thach, L., 1995. Using electronic mail to conduct survey research. *Educational Technology*, 35(2), pp. 27–31.

25

Terminologies

25.1 Pharmacy Research and Online Pharmacy Research Terminologies

The most common pharmacy research and online pharmacy research terminologies are the following:

Pharmacy Education Research

Any research related to pharmacy education issues such as teaching strategies, assessment methods, and others.

Pharmacy Practice Research

Any research related to pharmacy practice issues such as pharmacist care, pharmacists' prescribing, pharmacist-led services, and others.

Cross-Sectional Studies

These examine exposures and outcomes in populations at one point in time; they have no time sense (Strom, 2005).

Descriptive Studies

Studies that do not have control groups, namely case reports, case series, and analyses of secular trends. They contrast with analytic studies (Strom, 2005).

Case Control Study

Study that identifies a group of persons with the unintended drug effect of interest and a suitable comparison group of people without the unintended effect. The relationship of a drug to the drug event is examined by comparing the groups exhibiting and not exhibiting the drug event with regard to how frequently the drug is present (WHO, 2002).

Case control studies begin by identifying a sample of individuals with the outcome of interest (e.g., cancer or death) to serve as the cases and another sample without the outcome of interest to serve as the controls. Within both of these samples, the researcher then determines the exposure status of each individual (Harpe, 2011).

Clinical Trial

A systematic study on pharmaceutical products in human subjects (including patients and other volunteers) in order to discover or verify the effects of and/or identify any adverse reaction to investigational products, and/or to study the absorption, distribution, metabolism, and excretion of the products with the objective of ascertaining their efficacy and safety. Clinical trials are generally classified into Phases I to IV. Phase IV trials are studies performed after marketing of the pharmaceutical product. They are carried out on the basis of the product characteristics for which marketing authorization was granted and are normally in the form of post-marketing surveillance (WHO, 2002).

Cohort Study

A study that identifies defined populations and follows them forwards in time, examining their rates of disease. A cohort study generally identifies and compares exposed patients to unexposed patients or to patients who receive a different exposure (WHO, 2002).

Case Reports

Reports of the experience of single patients. As used in pharmacoepidemiology, a case report describes a single patient who was exposed to a drug and experiences a particular outcome, usually an adverse event (Strom, 2005).

Case Series

Reports of collections of patients, all of whom have a common exposure, examining what their clinical outcomes were. Alternatively, case series can be reports of patients who have a common disease, examining what their antecedent exposures were. No control group is present (Strom, 2005).

Ecological Studies

These examine trends in disease events over time or across different geographic locations and correlate them with trends in putative exposures, such as rates of drug utilization. The unit of observation is a subgroup or a population rather than individuals (WHO, 2002).

Quantitative Studies

Quantitative research is research that uses numerical analysis (GSU Library Research Guides).

Qualitative Studies

Research that derives data from observation, interviews, or verbal interactions and focuses on the meanings and interpretations of the participants (GSU Library Research Guides).

Retrospective Studies

Studies used to test etiologic hypotheses in which inferences about an exposure to putative causal factors are derived from data relating to characteristics of persons under study or to events or experiences in their past. The essential feature is that some of the persons under study have the disease or outcome of interest and their characteristics are compared with those of unaffected persons (GSU Library Research Guides).

Prospective Studies

Observation of a population for a sufficient number of persons over a sufficient number of years to generate incidence or mortality rates subsequent to the selection of the study group (GSU Library Research Guides).

Sample Size

The number of units (persons, animals, patients, specified circumstances, etc.) in a population to be studied. The sample size should be big enough to have a high likelihood of detecting a true difference between two groups (GSU Library Research Guides).

Bias

Any deviation of results or inferences from the truth, or processes leading to such deviation. Bias can result from several sources: one-sided or systematic variations in measurement from the true value (systematic error); flaws in study design; deviation of inferences, interpretations, or analyses based on flawed data or data collection; and so on. There is no sense of prejudice or subjectivity implied in the assessment of bias under these conditions (GSU Library Research Guides).

Pharmacoeconomics Research

"Pharmacoeconomics research identifies, measures, and compares the costs (ie, resources consumed) and consequences (ie, clinical, economic, humanistic) of pharmaceutical products and services. Within this framework are included the research methods related to cost-minimization, cost-effectiveness, cost–benefit, cost-of-illness, cost-utility, cost-consequences, and decision analysis, as well as quality-of-life and other humanistic assessments. In essence, pharmacoeconomic analysis uses tools for examining the impact (desirable, undesirable) of alternative drug therapies and other medical interventions" (Bootman et al., 1996).

Cost-Minimization Analysis

"When two or more interventions are evaluated and demonstrated or assumed to be equivalent in terms of a given outcome or consequence, costs associated with each intervention may be evaluated and compared. This typical cost analysis is referred to as cost-minimization analysis. An example of this type of investigation regarding drug therapy may be the evaluation of two generically equivalent drugs in which the outcome has been proven to be equal, although the acquisition and administration costs may be significantly different" (Bootman et al., 1996).

Cost-Benefit Analysis

"Cost-benefit analysis is a basic tool that can be used to improve the decision-making process in allocation of funds to healthcare programs. Although the general concept of cost-benefit analysis is not overly complicated, many technical considerations require a degree of explanation and interpretation to understand how it can be or has been applied. Cost-benefit analysis consists of identifying all of the benefits that accrue from the program or intervention and converting them into dollars in the year in which they will occur" (Bootman et al., 1996).

Cost-Effectiveness Analysis

"Cost-effectiveness analysis is a technique designed to assist a decision-maker in identifying a preferred choice among possible alternatives. Generally, cost-effectiveness is defined as a series of analytical and mathematical procedures that aid in the selection of a course of action from various alternative approaches. Cost-effectiveness analysis has been applied to health matters where the program's inputs can be readily measured in dollars, but the program's outputs are more appropriately stated in terms of health improvement created (eg, life-years extended, clinical cures)" (Bootman et al., 1996).

Cost-Utility Analysis

"Cost utility analysis is an economic tool in which the intervention consequence is measured in terms of quantity and quality of life. It is much the same as cost-effectiveness analysis, with the added dimension of a particular point of view, most often that of the patient" (Bootman et al., 1996).

Cost of Illness Analysis

"Cost of illness analysis is the determination of all costs of a particular disease, which include both direct and indirect costs. Since both costs were calculated, an economic evaluation for the disease can be performed successfully. It has been used for evaluating many diseases" (Bootman et al., 1996).

Literature Review

Generic term: published materials that provide examination of recent or current literature.

Can cover a wide range of subjects at various levels of completeness and comprehensiveness. May include research findings (Grant and Booth, 2009).

Traditional (Narrative) Literature Review

A generic review which identifies and reviews published literature on a topic, which may be broad. Typically employs a narrative approach to reporting the review findings. Can include a wide range of related subjects.

Rapid Review

Assesses what is known about an issue by using a systematic review method to search and appraise research and determine best practice (Grant and Booth, 2009).

Scoping Review

Assesses the potential scope of the research literature on a particular topic. Helps determine gaps in the research (Grant and Booth, 2009).

Critical Review

Aims to demonstrate that the writer has extensively researched the literature and critically evaluated its quality. Goes beyond mere description to include degree of analysis and conceptual innovation. Typically results in a hypothesis or model (Grant and Booth, 2009).

Systematic Review

A summary of the clinical literature. A systematic review is a critical assessment and evaluation of all research studies that address a particular clinical issue. The researchers use an organized method of locating, assembling, and evaluating a body of literature on a particular topic using a set of specific criteria. A systematic review typically includes a description of the findings of the collection of research studies. The systematic review may also include a quantitative pooling of data, called a meta-analysis (GSU Library Research Guides).

Systematic Search and Review

Combines the strengths of a critical review with a comprehensive search process. Typically addresses broad questions (Grant and Booth, 2009).

Umbrella Review

Specifically refers to a review compiling evidence from multiple reviews into one accessible and usable document. Focuses on a broad condition or problem for which there are competing interventions and highlights reviews that address these interventions and their results.

Mixed Methods Review

Refers to any combination of methods where one significant component is a literature review (usually systematic). Within a review context it refers to a combination of review approaches, for example, combining quantitative with qualitative research or outcome with process studies (Grant and Booth, 2009).

Meta-Analysis

Works consisting of studies using a quantitative method of combining the results of independent studies (usually drawn from the published literature) and synthesizing summaries and conclusions which may be used to evaluate therapeutic effectiveness, plan new studies, and so on. It is simply a way of combining data from many different research studies. A meta-analysis is a statistical process that combines the findings from individual studies (GSU Library Research Guides).

25.2 Conclusion

This chapter has discussed the various terminologies used in pharmacy and online pharmacy research. This chapter included the terminologies related to all study design in both pharmacy and online pharmacy research.

References

Bootman, J.L., Townsend, R.J. and McGhan, W.F., 1996. *Introduction to pharmacoeconomics*. Principles of pharmacoeconomics, 2.

Grant, M.J. and Booth, A., 2009. A typology of reviews: An analysis of 14 review types and associated methodologies. *Health Information & Libraries Journal*, 26(2), pp. 91–108. GSU Library Research Guides. https://research.library.gsu.edu/c.php?g=115595&p=755213

Harpe, S., 2011. *Study designs for pharmacoepidemiology* (pp. 39–54). McGraw-Hill Companies.

Strom, B.L., 2005. *Pharmacoepidemiology*. 4th ed. John Wiley & Sons.

World Health Organization., 2002. *The importance of pharmacovigilance*. World Health Organization, 2004. *Pharmacovigilance: Ensuring the safe use of medicines* (No. WHO/EDM/2004.8). World Health Organization.

26

Research Methods and Methodology

26.1 Importance of Study Methods and Methodology

To select a suitable study design to be able to answer the study questions, study design is very important, and research should be valid and reliable to address the study questions. Without appropriate study design, the study questions will not be addressed and the study objectives will not be achieved. Researchers can read similar published studies and adapt their study design or take advice from their colleagues/professionals/experts before conducting the study about which study design he/she should select for his/her study. There are several different schemes for classifying study designs as discussed next (Dawson and Trapp, 2001; Röhrig et al., 2009; Grant and Booth, 2009).

26.2 Research Method and Methodology Issues

26.2.1 Classification Based on Data Sources: Primary versus Secondary Research

Primary research includes the following:

Basic research

Clinical research

Epidemiological research

Secondary research includes the following:

Review: systematic, narrative, scoping, umbrella, qualitative systematic review/qualitative evidence synthesis

Meta-analyses

26.2.2 Classification Based on Outcome Exposure

26.2.2.1 Retrospective Study Design

From outcome to exposure, in other words, what happened in the past.

26.2.2.2 Prospective Study Design

From exposure to outcome, in other words, what will happen in the future.

26.2.3 Classification Based on Research Purpose

26.2.3.1 Descriptive Studies

Design to describe the practice, situation, data, and others. For example, describing the pharmacovigilance system in your country.

26.2.3.2 Analytic Studies

Design to examine/investigate the practice, data, and others.

26.2.4 Classification Based on Data Collection Type

26.2.4.1 Quantitative Studies

In this study design, researchers use numerical analysis, such as a survey with Likert scale questions.

26.2.4.2 Qualitative Studies

In this study design, researchers derive data from observation, interviews, or verbal interactions and focus on the meanings and interpretations of the participants.

26.2.4.3 Mixed Method Studies

In this study design, researchers combine both quantitative and qualitative study designs.

26.2.4.4 Simulation Studies

In this study design, researchers will use an actor, a simulation patient, to investigate the practice. Example: Quality of dispensing medication at community pharmacy. This study will use simulation study design. Two or more simulation patients will visit a number of pharmacies and evaluate their dispensing practice quality by using an evaluation checklist.

26.2.5 Classification of Analytic Studies

26.2.5.1 Non-Experimental (Observational Studies)

- Cohort (retrospective and prospective)
- Case control
- Cross-sectional
- Ecological

26.2.5.2 Quasi-Experimental

A quasi-experiment is an empirical interventional study used to estimate the causal impact of an intervention on a target population without randomization. The investigator lacks full control over the intervention.

26.2.5.3 Experimental (Intervention Studies)

- Controlled trials: randomized, non-randomized
- Uncontrolled trials

26.2.6 Classification of Pharmacoeconomic Studies

26.2.6.1 Cost-Minimization Analysis

"When two or more interventions are evaluated and demonstrated or assumed to be equivalent in terms of a given outcome or consequence, costs associated with each intervention may be evaluated and compared. This typical cost analysis is referred to as cost-minimization analysis. An example of this type of investigation regarding drug therapy may be the evaluation of two generically equivalent drugs in which the outcome has been proven to be equal, although the acquisition and administration costs may be significantly different" (Bootman et al., 1996).

26.2.6.2 Cost-Benefit Analysis

"Cost-benefit analysis is a basic tool that can be used to improve the decision-making process in allocation of funds to healthcare programs. Although the general concept of cost-benefit analysis is not overly complicated, many technical considerations require a degree of explanation and interpretation to understand how it can be or has been applied. Cost-benefit analysis consists of identifying all of the benefits that accrue from the program or intervention and converting them into dollars in the year in which they will occur" (Bootman et al., 1996).

26.2.6.3 Cost-Effectiveness Analysis

"Cost-effectiveness analysis is a technique designed to assist a decision-maker in identifying a preferred choice among possible alternatives. Generally, cost-effectiveness is defined as a series of analytical and mathematical procedures that aid in the selection of a course of action from various alternative approaches. Cost-effectiveness analysis has been applied to health matters where the program's inputs can be readily measured in dollars, but the program's outputs are more appropriately stated in terms of health improvement created (eg, life-years extended, clinical cures)" (Bootman et al., 1996).

26.2.6.4 Cost-Utility Analysis

"Cost utility analysis is an economic tool in which the intervention consequence is measured in terms of quantity and quality of life. It is much the same as cost-effectiveness analysis, with the added dimension of a particular point of view, most often that of the patient" (Bootman et al., 1996).

26.2.6.5 Cost of Illness Analysis

"Cost of illness analysis is the determination of all costs of a particular disease, which include both direct and indirect costs. Since both costs were calculated, an economic evaluation for the disease can be performed successfully. It has been used for evaluating many diseases" (Bootman et al., 1996).

26.3 Study Methods and Methods for Online Research

Pharmacy researchers can select/use many study methods and methodologies for their research such as the following:

Literature review

Online survey

Online interviews

Online focus groups

Online qualitative research

Web-based experiments

Online clinical trials

Online simulation

26.4 Conclusion

This chapter has discussed the most common research methods and methodologies in pharmacy and online research. To select a suitable study design to be able to answer the study questions, study design is very important, and research should be valid and reliable to address the study questions. Without appropriate study design, the study questions will not be addressed and the study objectives will not be achieved.

References

Bootman, J.L., Townsend, R.J. and McGhan, W.F., 1996. Introduction to pharmacoeconomics. In *Principles of pharmacoeconomics*, 2.

Dawson, B. and Trapp, R.G., 2001. Study designs in medical research. In *Basic and clinical biostatistics* (pp. 7–23). McGraw-Hill.

Grant, M.J. and Booth, A., 2009. A typology of reviews: An analysis of 14 review types and associated methodologies. *Health Information & Libraries Journal*, 26(2), pp. 91–108.

Röhrig, B., Du Prel, J.B., Wachtlin, D. and Blettner, M., 2009. Types of study in medical research: Part 3 of a series on evaluation of scientific publications. *Deutsches Arzteblatt International*, 106(15), p. 262.

27

Tips for Implementation

27.1 Background

Research is very important for pharmacy educators, students, and professionals. Being a researcher and able to conduct research is a very important competency for pharmacy students and the future pharmacist. Pharmacy schools/departments aim to graduate pharmacists with the ability to conduct research; therefore, there is a graduation project in the majority of pharmacy programs worldwide. In order to conduct good research, researchers should plan it appropriately from the beginning to avoid any delay or change in the research (Al-Worafi, 2020). A good plan includes the steps outlined next.

27.1.1 Select the Area

At this step, the researcher needs to select the area, which could be related to the pharmacy education or pharmacy practice. Moreover, which subarea could be related to the pharmacy education or pharmacy practice. For example: plan to conduct research in pharmacy education, in teaching strategies; plan to conduct research in pharmacy practice, in patient education and counseling.

27.1.2 Select the Topic

At this step you need to define your topic, or the title of your research. The topic should be consistent with the study objectives/questions.

27.1.3 Justify the Need for Your Study

Why are you going to conduct your study? Explain why you need to conduct your study. Is there a gap in the literature about the selected topic? Why is the selected topic important? Is the selected topic having an impact on pharmacy education and practice?

DOI: 10.1201/9781003230458-30

27.1.4 Write the Background about the Selected Topic

To be successful in your research, you should read about your topic's basics and background. This will give you as well as the readers/reviewers an overview about your topic's basics, importance, and gaps in the literature. To search about your topic, select the appropriate keywords to make the search easy for you.

27.1.5 Define the Objectives/Questions

At this step, you need to define the aim and objective/questions of the study.

27.1.6 Define the Study Hypothesis

At this step, you need to define the null as well as alternative hypothesis based on the study questions. The null hypothesis is the opposite of the alternative hypothesis.

27.1.7 Describe the Significance of Your Study

Why this study important?

It Will help in . . .

It will explore . . .

It will identify . . .

It will investigate . . .

It will help policy makers, researchers, and health care professionals . . .

It will improve . . .

27.1.8 Define the Expected Outcomes

What do you expect to find at the end of this study?

27.1.9 Approval of the Study

For studies involving human subjects or animals you need to get approval of your study before conduction from the ethical committees, either in the University or Ministry of Health or others.

27.1.10 Determine the Appropriate Methodology

To select a suitable study design to be able to answer the study questions, study design is very important, and research should be valid and reliable to address the study questions. Without appropriate study design, the study questions will not be addressed and the study objectives will not be achieved. Researchers can read similar published studies and adapt their study design or take advice from colleagues/professionals/experts before conducting the study about which study design he/she should select for his/her study. The following examples can be used for pharmacy research and online research:

Review: Systematic, narrative, scoping, umbrella, qualitative systematic review/qualitative evidence synthesis.

Meta-analyses.

Retrospective study design: From outcome to exposure, in other words, what happened in the past.

Prospective study design: From exposure to outcome, in other words, what will happen in the future.

Descriptive studies: Design to describe the practice, situation, data, and others such as describing the pharmacovigilance system in your country.

Analytic studies: Design to examine/investigate the practice, data, others.

Quantitative studies: In this study design, researchers use numerical analyses, such as a survey with Likert scale questions.

Qualitative Studies: In this study design, researchers derive data from observation, interviews, or verbal interactions and focus on the meanings and interpretations of the participants.

Mixed method studies: In this study design, researchers combine both the quantitative and qualitative study designs.

Simulation studies: In this study design, researchers will use an actor, a simulation patient, to investigate the practice. Example: Quality of dispensing medication at community pharmacies. This study will use the simulation study design. Two or more simulation patients will visit a number of pharmacies and evaluate their dispensing practice quality by using an evaluation checklist.

27.1.10.1 Classification of Analytic Studies

Non-Experimental (Observational Studies)

- Cohort (retrospective and prospective)
- Case control
- Cross-sectional
- Ecological

Quasi-Experimental

A quasi-experiment is an empirical interventional study used to estimate the causal impact of an intervention on a target population without randomization. The investigator lacks full control over the intervention.

Experimental (Intervention Studies)

- Controlled trials: randomized, non-randomized
- Uncontrolled trials

27.1.10.2 Study Tool

Develop/adapt a data collection form; surveys; simulation scenarios using a validated checklist, interview guides, and other tools. Be sure that your study tool is valid and reliable.

27.1.10.3 Sampling Procedure

Sample Size

There are many methods for calculating the sample size and it depends on the type of study.

There are also many online websites for sample size calculations, but they can't be used for all types of studies. You may read the previous/similar work as a guide or ask colleagues to help you in selecting the appropriate sample size calculation method.

Sampling Method

There are many sampling methods as follows:

- Convenience sampling
- Quota sampling
- Snowball sampling
- Systematic sampling
- Simple random sampling
- Stratified sampling
- Clustered sampling

27.1.10.4 Criteria for Subject's Selection

The subject's selection in this study will be based on inclusion and exclusion criteria.

Inclusion Criteria
Example: Adult patients.

Exclusion Criteria

Example: Pediatric patients.

27.1.10.5 Data Analysis Procedure/Statistical Analysis

There are many statistical programs available to analyze your data such as: SPSS, STATA, MINITAB, and so on. You should select the appropriate tests to analyze your data. You may ask statistical analysis experts to analyze your data.

27.1.11 Results

What are the findings of your study?

Present the data as: tables, figures, or text based on your data and preference and the journal's or institution's requirements. Present it in a simple and easy way for readers to understand it.

27.1.12 Discussion

What do the results mean?

What are the similarities and differences between your finding and those of other studies worldwide?

27.1.13 Conclusion

State the conclusions and implications of the results; also, describe the limitations and recommendations for further studies.

27.1.14 References Style

There are many styles of references. The most common styles include Harvard, Oxford, APA, Chicago, and Vancouver (numeric). There are many programs to generate (write) your references such as EndNote. There are also many websites that can generate the references for free. You can also cite articles directly from Google Scholar.

27.2 Tips for Publishing Research

Publishing research in high-quality journals is the key performance indicator for your research. Here are a few tips that could help you in publishing your research:

Select the appropriate journal for your work.

Write and structure your manuscript based on journal guidelines/instructions.

Write your title and abstract as well as the whole manuscript in an attractive and clear way.

Send your article for English proofreading.

Be familiar with the common reasons for rejecting manuscripts such as importance of topics; study design; presentation of manuscript; and the presentation/discussion of your results, to avoid rejection.

27.3 Barriers to Implementing Online Research

There are many barriers to implementing online research in pharmacy as follows:

Lack of funds.

Lack of facilities.

Lack of technologies.

Attitude towards online research.

Knowledge towards online research.

Lack of training.

Lack of motivation.

27.4 Conclusion

This chapter has discussed online pharmacy research. This chapter includes the best practices in implementing online research from the plan till the publication of research. Furthermore, it includes the barriers to implementing online research.

Reference

Al-Worafi, Y.M., 2020. Medications safety research issues. In *Drug safety in developing countries* (pp. 213–227). Academic Press.

28

Quality of Online Research

28.1 Importance of Quality in Pharmacy Education and Practice Research

Conducting quality research in pharmacy education and practice is very important and provides reliable and accurate, valid results that could help to identify pharmacy education and pharmacy practice challenges and design recommendation to overcome them. It could improve pharmacy education and pharmacy practice. Quality research will lead to evidence-based practice, which is essential to improving the quality of health care, patient care, and pharmacist care and services. Pharmacy researchers should take into consideration the quality of research from the first step, which is the research plan. However, also pharmacy schools, hospitals, and organizations should ensure and supervise the quality of research.

28.2 Quality of Research Indicators and Criteria

Pharmacy researchers can ask themselves the following questions to ensure the quality of their research.

28.2.1 Appropriateness of Research/Study Topic and Title

What is the title of the research?

28.2.2 Is the Topic Consistent with the Study Objectives/Questions?

Then, pharmacy researchers can modify the title to be consistent with the research/study objectives/questions.

DOI: 10.1201/9781003230458-31

28.2.3 Justify the Need for the Research/Study

Why are you going to conduct your study?

Explain why you need to conduct your study. Is there a gap in the literature about the selected topic?

Why is the selected topic important?

Does the selected topic have an impact on pharmacy education and practice?

28.2.4 Appropriateness of Background about the Selected Topic

Is the background appropriate and does it give an overview about your topic's basics, importance, and gaps in the literature?

Does the literature review cover enough?

Does the review include recent literature?

Does the content of the review relate to the research problem, objectives, and hypothesis?

28.2.5 Appropriateness of the Research/Study Objectives and Questions

Are the research/study objective and questions appropriate?

28.2.6 Appropriateness of the Research/Study Hypothesis

Is the research/study hypothesis appropriate?

28.2.7 Appropriateness of the Significance of Research/Study and Expected Outcomes

Is the research/study significant?

Does the research/study have an expected outcome?

28.2.8 Appropriateness of the Research/Study Methods and Methodology

Is the study design appropriate?

Are the study tools valid and reliable?

Are the study methods explained with full and all details?

Is the sample size appropriate?

Is the sampling technique appropriate?

Who are the subjects?

What were the inclusion/exclusion criteria for the study?

How were subjects recruited/selected?

How were the data analyzed?

Are the statistical tests appropriate and valid?

28.2.9 Appropriateness of the Research/Study Results

What are the findings of your study?

Are the results for each objective and question achieved?

Are the results presented in a clear and understandable way?

Are the results significant?

28.2.10 Appropriateness of the Research/Study Discussion

What do the results mean?

What are the similarities and differences between your findings and those of other studies worldwide?

Are the interpretations consistent with the results?

Are the conclusions accurate and relevant to the research problem?

Are the recommendations appropriate?

Are study limitations addressed?

28.3 Peer Review and Quality of Research

Peer reviews play an important role in the quality of research. Pharmacy researchers can meet with colleagues and ask them to evaluate the research proposal or papers. There are many forms for this purpose such as:

Appropriateness of Study Title

Very poor	Poor	Average	Very Good	Outstanding
1	2	3	4	5

Appropriateness of Study Background

Very poor	Poor	Average	Very Good	Outstanding
1	2	3	4	5

Appropriateness of the Rationality of the Study

Very poor	Poor	Average	Very Good	Outstanding
1	2	3	4	5

Appropriateness of the Study Objectives (or Hypothesis)

Very poor	Poor	Average	Very Good	Outstanding
1	2	3	4	5

Appropriateness of the Study Design (or Methods)

Very poor	Poor	Average	Very Good	Outstanding
1	2	3	4	5

Appropriateness of the Subject Selection and Sample Size Calculation

Very poor	Poor	Average	Very Good	Outstanding
1	2	3	4	5

Appropriateness of Outcome Measurement

Very poor	Poor	Average	Very Good	Outstanding
1	2	3	4	5

Appropriateness of Analysis (or Statistical Analysis)

Very poor	Poor	Average	Very Good	Outstanding
1	2	3	4	5

Appropriateness of Results (Are Results for Each Objective and Outcome Achieved?)

Very poor	Poor	Average	Very Good	Outstanding
1	2	3	4	5

Appropriateness of Discussion (Are Study Findings Explained and Compared with Other Studies?)

Very poor	Poor	Average	Very Good	Outstanding
1	2	3	4	5

Appropriateness of Conclusion (Summarize Study Findings, Strength, Limitation, Future Recommendations, and Application of Study)

Very poor	Poor	Average	Very Good	Outstanding
1	2	3	4	5

28.4 Quality of Research Papers

There are many tools for assessing the quality of research papers based on study design. Examples of these tools are as follows (BMJ, 2021):

Randomized Controlled Trials (RCTs)

CONSORT guidelines, flowcharts, and structured abstract checklists.

Systematic Reviews and Meta-Analyses

PRISMA guidelines, flowcharts, and structured abstract checklists.

Observational Studies in Epidemiology

STROBE guidelines (also refer to RECORD for observational studies using routinely collected health data) and MOOSE guidelines.

Diagnostic Accuracy Studies

STARD guidelines.

Quality Improvement Studies

SQUIRE guidelines.

Multivariate Prediction Models

TRIPOD guidelines.

Economic Evaluation Studies

CHEERS guidelines.

Animal Pre-Clinical Studies

ARRIVE guidelines.

Web-Based Surveys

CHERRIES guidelines.

28.5 Conclusion

This chapter has discussed the quality of research and online research issues and the quality of research articles. Conducting quality research in pharmacy education and practice is very important and provides reliable and accurate, valid results that could help to identify pharmacy education and pharmacy practice challenges and design recommendations to overcome them. It could improve the practice of pharmacy education and pharmacy practice. Quality research will lead to evidence-based practice, which is essential to improving the quality of health care, patient care, and pharmacist care and services.

Reference

BMJ., 2021. *Reporting guidelines*. Available at: https://authors.bmj.com/before-you-submit/reporting-guidelines/

29

Facilitators and Barriers

29.1 Facilitators

Pharmacy education and practice research is very important to educators, students, professionals, researchers, and policy makers. The facilitators for conducting pharmacy research and online pharmacy research are:

Funding

Funding is one of the most important facilitators for conducting pharmacy research and online pharmacy research.

Research Skills/Competencies

Research skills/competencies are one of the most important facilitators for conducting pharmacy research and online pharmacy research. Therefore, pharmacy programs aim to graduate pharmacists with the essential knowledge and skills to be good researchers.

Interest

Interest is very important to conducting pharmacy research and online pharmacy research.

Attitude towards the Importance and Benefits of Research

The attitude of pharmacy researchers towards the research and its importance and benefits is very important to conducting pharmacy research and online pharmacy research.

Time

Allocating time for research is very important to conducting pharmacy research and online pharmacy research.

DOI: 10.1201/9781003230458-32

Communication

Communication is the key to success in any work. Effective communication skills are very important and required to conduct pharmacy research and online pharmacy research.

Teamwork

Teamwork is very important and is required to conduct pharmacy research and online pharmacy research.

Collaboration

Collaboration is very important and is required to conduct pharmacy research and online pharmacy research.

Suitable and Effective Technologies and Tools

Suitable and effective technologies and tools are the keys to success in online pharmacy research. Pharmacy researchers should use and adapt the most effective technologies and tools for online patient care such as the following:

Internet

The internet plays a very important and vital role in conducting pharmacy research and online pharmacy research.

Computers and Laptops

Using computers and laptops is very important and essential for effective pharmacy research and online pharmacy research.

Smartphones, Tablets, and Net Books

Using smartphones, tablets, and net books is very important and essential for effective pharmacy research and online pharmacy research.

Webinar and Video Conferencing Platforms

Webinar and video conferencing platforms are very important in pharmacy research and online pharmacy research. Microsoft Teams, Cisco WebEx Teams, Google Meet, and Zoom are very important for the success of online pharmacy research. They include free versions that are great for light uses, short conference calls, and light file sharing.

WhatsApp

WhatsApp can be used in pharmacy research and online pharmacy research as an effective communication tool between pharmacy researchers and participants as well as among researchers.

Wearable Technologies

Wearable technologies can be used in pharmacy research and online pharmacy research.

29.2 Barriers to Pharmacy Research and Online Pharmacy Research

29.2.1 Lack of Funding

Lack of funding is one of the most important barriers to conducting pharmacy research and online pharmacy research, especially in many low-income and middle-income countries; therefore, collaboration of pharmaceutical companies/industries, organizations/international organizations, policy makers, and universities to find a way to support research is very important and highly recommended.

29.2.2 Lack of Research Infrastructure

Lack of infrastructure to conduct pharmacy research and online pharmacy research is a major barrier to conducting pharmacy research and online pharmacy research, especially in many low-income and middle-income countries; therefore, collaboration of pharmaceutical companies/industries, organizations/international organizations, policy makers, and universities to find a way to support research is very important and highly recommended.

29.2.3 Inadequate Research Skills/Competencies

Inadequate research skills/competencies are major barriers to conducting pharmacy research and online pharmacy research. Therefore, improving research skills/competencies through continuous professional development (CPD) is very important and highly recommended.

29.2.4 Lack of Interest and Motivation

Lack of interest and motivation are major barriers to conducting pharmacy research and online pharmacy research. Increasing awareness about the importance of conducting pharmacy research and online pharmacy research could overcome this barrier.

29.2.5 Negative Attitude towards Research

Negative attitudes among pharmacy researchers is a major barrier to conducting pharmacy research and online pharmacy research. Increasing awareness about the importance of conducting pharmacy research and online pharmacy research could overcome this barrier.

29.2.6 Lack of Time

Lack of time among pharmacy researchers is a major barrier to conducting pharmacy research and online pharmacy research. Allocating time for pharmacy researchers could overcome this barrier.

29.2.7 Poor Communication

Poor communication among pharmacy researchers is a major barrier to conducting pharmacy research and online pharmacy research. Improving communication skills through continuous professional development (CPD) is very important and highly recommended to overcome this barrier.

29.2.8 Lack of Research Facilities

Lack of research facilities is a major barrier to conducting pharmacy research and online pharmacy research; therefore, policy makers, universities, pharmaceutical companies, and organizations should collaborate to overcome this barrier.

29.2.9 Lack of Collaboration

Lack of collaboration among pharmacy researchers and other health care professionals/researchers is a major barrier to conducting pharmacy research and online pharmacy research. Increasing awareness about the importance of collaboration could overcome this barrier.

29.2.10 Ethical

The absence of professional ethical committees is a barrier to conducting pharmacy research and online pharmacy research in many countries; therefore, adapting the international guidelines such as the World Medical Association Declaration of Helsinki, and taking experience from developed countries could help to overcome this barrier.

29.2.11 Lack/Absence of Suitable and Effective Technologies and Tools

Lack/absence of suitable and effective technologies and tools are major barriers to conducting pharmacy research and online pharmacy research, especially among researchers and participants in low-income and middle-income countries. Suitable and effective technologies and tools are the keys to success in pharmacy research and online pharmacy research. Pharmacy researchers should use and adapt the most effective technologies and tools for the online research as follows:

Internet

The internet plays a very important and vital role in conducting pharmacy research and online pharmacy research.

Computers and Laptops

Using computers and laptops is very important and essential for effective pharmacy research and online pharmacy research.

Smartphones, Tablets, and Net Books

Using smartphones, tablets, and net books is very important and essential for effective pharmacy research and online pharmacy research.

Webinar and Video Conferencing Platforms

Webinar and video conferencing platforms are very important in pharmacy research and online pharmacy research. Microsoft Teams, Cisco WebEx Teams, Google Meet, and Zoom are very important for the success of online pharmacy research. They include free versions that are great for light uses, short conference calls, and light file sharing.

WhatsApp

WhatsApp can be used in pharmacy research and online pharmacy research as an effective communication tool between pharmacy researchers and participants as well as among researchers.

Wearable Technologies

Wearable technologies can be used in pharmacy research and online pharmacy research.

29.3 Conclusion

This chapter has discussed the facilitators and barriers for online research. Pharmacy education and practice research are very important to educators, students, professionals, researchers, and policy makers.

Section 4

Pharmacy Education Teaching Strategies

30

Pharmacy Education: Learning Styles

30.1 Definition

Learning styles can be defined as the manner in which individuals choose to or are inclined to approach a learning situation (Cassidy, 2004).

30.2 Learning Styles Frameworks

Learning styles frameworks in pharmacy and health education include examples such as the following (Childs-Kean et al., 2020):

30.2.1 The VARK Learning Style

The acronym VARK stands for Visual, Aural, Read/write, and Kinesthetic sensory modalities that are used for learning information (Vark Learn Limited, 2021).

Visual (V)

This preference includes the depiction of information in maps, spider diagrams, charts, graphs, flow charts, labelled diagrams, and all the symbolic arrows, circles, hierarchies, and other devices that people use to represent what could have been presented in words. This mode could have been called Graphic (G) as that better explains what it covers. It does NOT include still pictures or photographs of reality, movies, videos, or PowerPoint. It does include designs, whitespace, patterns, shapes, and the different formats that are used to highlight and convey information. When a whiteboard is used to draw a diagram with meaningful symbols for the relationship between different things, that will be helpful for those with a Visual preference. It must be more than mere words in boxes, which would be more helpful to those who have a Read/write preference.

Aural/Auditory (A)

This perceptual mode describes a preference for information that is "heard or spoken." Learners who have this as their main preference report that they learn best from lectures, group discussion, radio, email, using mobile phones, speaking, web-chat, and talking things through. Email is included here because although it is text and could be included in the Read/write category (later), it is often written in chat style with abbreviations, colloquial terms, slang, and nonformal language. The Aural preference includes talking out loud as well as talking to oneself. Often people with this preference want to sort things out by speaking first, rather than sorting out their ideas and then speaking. They may say again what has already been said, or ask an obvious and previously answered question. They have a need to say it themselves and they learn through saying it their way.

Read/write (R)

This preference is for information displayed as words. Not surprisingly, many teachers and students have a strong preference for this mode. Being able to write well and read widely are attributes sought by employers of graduates. This preference emphasizes text-based input and output—reading and writing in all its forms but especially manuals, reports, essays, and assignments. People who prefer this modality are often addicted to PowerPoint, the internet, lists, diaries, dictionaries, thesauri, quotations, and words, words, words. Note that most PowerPoint presentations and the internet, Google, and Wikipedia are essentially suited to those with this preference as there is seldom an auditory channel or a presentation that uses visual symbols.

Kinesthetic (K)

By definition, this modality refers to the "perceptual preference related to the use of experience and practice (simulated or real)." Although such an experience may invoke other modalities, the key is that people who prefer this mode are connected to reality, "either through concrete personal experiences, examples, practice or simulation." It includes demonstrations, simulations, videos and movies of "real" things, as well as case studies, practice, and applications. The key is the real or concrete nature of the example. If it can be grasped, held, tasted, or felt it will probably be included. People with this as a strong preference learn from the experience of doing something and they value their own background of experiences and less so, the experiences of others. It is possible to write or speak kinesthetically if the topic is strongly based in reality. An assignment that requires the details of who will do what and when is suited to those with this preference, as is a case study or a working example of what is intended or proposed.

30.2.2 Kolb Learning Style Inventory

Kolb's experiential learning style theory is typically represented by a four-stage learning cycle in which the learner "touches all the bases" (McLeod, 2017):

1. Concrete Experience—a new experience or situation is encountered, or a reinterpretation of existing experience.
2. Reflective Observation of the New Experience—of particular importance are any inconsistencies between experience and understanding.
3. Abstract Conceptualization—reflection gives rise to a new idea, or a modification of an existing abstract concept (the person has learned from their experience).
4. Active Experimentation—the learner applies their idea(s) to the world around them to see what happens.

Descriptions of the Four Kolb Learning Styles (McLeod, 2017)

Diverging (Feeling and Watching)

These people are able to look at things from different perspectives. They are sensitive. They prefer to watch rather than do, tending to gather information and use imagination to solve problems. They are best at viewing concrete situations from several different viewpoints.

Assimilating (Watching and Thinking)

The assimilating learning preference involves a concise, logical approach. Ideas and concepts are more important than people. These people require good, clear explanation rather than a practical opportunity. They excel at understanding wide-ranging information and organizing it in a clear, logical format. People with an assimilating learning style are less focused on people and more interested in ideas and abstract concepts. People with this style are more attracted to logically sound theories than approaches based on practical value.

Converging (Doing and Thinking)

People with a converging learning style can solve problems and will use their learning to find solutions to practical issues. They prefer technical tasks, and are less concerned with people and interpersonal aspects. People with a converging learning style are best at finding practical uses for ideas and theories. They can solve problems and make decisions by finding solutions to questions and problems. People with a converging learning style are more attracted to technical tasks and problems than social or interpersonal issues.

Accommodating (Doing and Feeling)

The accommodating learning style is "hands-on," and relies on intuition rather than logic. These people use other people's analysis, and prefer to take a practical, experiential approach. They are attracted to new challenges and experiences, and to carrying out plans.

30.2.3 Honey and Mumford Learning Style Questionnaire

This learning style consists of the following (Honey et al., 1992):

Activist: Likes to take action.

Reflector: Likes to think before they act.

Theorist: Likes logic and likes to see both details and the overall picture.

Pragmatist: Likes practicality and experimenting.

30.2.4 Pharmacist Inventory of Learning Styles

This learning style consists of the following (Austin, 2004):

Enactors: Doing and unstructured environment.

Producers: Reflecting and structured environment.

Directors: Doing and structured environment.

Creators: Reflecting and unstructured environment.

30.2.5 Gregorc Style Delineator

This learning style consists of the following (Coffield et al., 2004):

Concrete Sequential: Order and practicality.

Abstract Sequential: Logic and rationales.

Abstract Random: Spontaneity and emotions.

Concrete Random: Originality and independence.

30.2.6 Grasha-Reichmann Student Learning Style Scale

This learning style consists of the following (Novak et al., 2006):

Independent: Solo work.

Avoidant: Avoids participation.

Collaborative: Works well with peers and faculty.

Dependent: Works within specific guidelines.

Competitive: Competition and winning.

Participant: Joins all available learning activities.

30.2.7 4MAT Learning Style

4MAT is a model for creating more dynamic and engaging learning. It is a framework for learning that helps educators deliver information in more dynamic and engaging ways.

While traditional instruction may focus primarily on facts and information (What?) the 4MAT model encourages a broader array of questions to elicit much higher levels of student understanding and involvement (aboutlearning.com). This learning style consists of the following (aboutlearning.com):

Why?

We want to understand meaning and purpose, and the instructor's role is to make connections between the material and the learners, to engage their attention.

What?

Only when we are satisfied about relevance are we ready to know *"What?"* At this stage, the trainer provides information and satisfies our desire for facts, structure, and theory.

These first two phases represent instructor-led learning. Now the learner takes over.

How?

Once we have the knowledge, we ask *"How?"* and we want to understand how we can apply our new insights to the real world. We focus on problems and how we can use our learning to solve them.

What if?

Finally, we want to try it out, so we ask questions like *"What if?"* "What else?" or "What next?" This is when we engage in active experimentation, trial and error, pushing at the boundaries—learning by doing.

30.3 Conclusion

This chapter has discussed the learning styles in pharmacy education. This chapter includes the different learning styles such as: VARK learning style; the Kolb Learning Style Inventory; and 4MAT learning style.

References

Austin, Z., 2004. Development and validation of the Pharmacists' Inventory of Learning Styles (PILS). *American Journal of Pharmaceutical Education*, 68(2).

Cassidy, S., 2004. Learning styles: An overview of theories, models, and measures. *Educational Psychology*, 24(4), pp. 419–444.

Childs-Kean, L., Edwards, M. and Smith, M.D., 2020. Use of learning style frameworks in health science education. *American Journal of Pharmaceutical Education*, 84(7).

Coffield, F., Moseley, D., Hall, E., Ecclestone, K., Coffield, F., Moseley, D., Hall, E. and Ecclestone, K., 2004. *Learning styles and pedagogy in post-16 learning: A systematic and critical review*. Available at: https://elearningindustry.com/critical-analysis-of-learning-styles-pedagogy-post-16-learning

Honey, P. and Mumford, A., 1992.*The manual of learning styles*. 3rd ed. Honey Press. https://aboutlearning.com/about-us/4mat-overview/

Vark Learn Limited., 2021. *The VARK modalities*. Available at: https://vark-learn.com/introduction-to-vark/the-vark-modalities/

McLeod, S., 2017. Kolb's learning styles and experiential learning cycle. *Simply Psychology*, 5.

Novak, S., Shah, S., Wilson, J.P., Lawson, K.A. and Salzman, R.D., 2006. Pharmacy students' learning styles before and after a problem-based learning experience. *American Journal of Pharmaceutical Education*, 70(4).

31

Traditional and Active Strategies

31.1 Traditional Teaching Strategies

Traditional teaching strategies were lecture-based and teacher-based, where the educator delivered the lecture and the students listened. Despite the revolution in teaching strategies and the many active teaching strategies that have been developed and implemented successfully in pharmacy education around the world, lecture-based teaching strategy remains a backbone of pharmacy education and the preferred teaching strategy among pharmacy educators in many countries.

31.2 Active Teaching Strategies

There are many effective teaching strategies that can be used in pharmacy education such as the following (McCoy et al., 2018; Al-Meman et al., 2014; Stewart et al., 2011):

Audience Response System/Clickers

Use of remote control devices by students to anonymously respond to multiple-choice questions posed by the instructor; can be integrated into traditional lectures, often termed "active lecture." Audience response: Individual students respond to application of skill questions via an audience response system.

Discussion-Based Learning, Including Deliberative Discussion

Use of communication among learners (both synchronous and asynchronous) as a teaching modality; can be used with other strategies such as case studies.

Interactive-Spaced Education

Use of repetition of content at spaced intervals combined with testing of that content; developed and used heavily within the context of medical education.

Interactive Web-Based Learning

Use of web-based modules to deliver content and assess student understanding in an interactive format.

Patient Simulation

Use of human patient simulators in a laboratory environment to teach providers to respond to a variety of physiological emergencies and situations.

Process-Oriented Guided Inquiry Learning (POGIL)/ Discovery Learning

Use of exercises specifically designed to lead teams of students through the stages of exploring data, developing concepts based on that data, and applying the concepts.

Problem-Based Learning (PBL), Including Case-Based Learning

Use of cases or problem sets meant to be explored in self-managed teams of students (with a facilitator); PBL sessions precede any discussion of content by the instructor.

Team-Based Learning (TBL)

Use of small student groups to facilitate discussion, case study exploration, or other aspects of content; preparation is required in advance and content is integrated throughout the class by the facilitator (expert).

Traditional Laboratory Experiences

Use of traditional laboratory and benchtop experiences to provide hands-on learning experiences.

Vodcast + Pause Activities

A video podcast with pause activities, appended exercises, or practice questions.

Vodcast + Hyperlinks

A video podcast with no pause activities but that includes hyperlinks to external or Web media for enrichment.

Interactive Vodcast

A vodcast that requires students to physically click through questions or interactivities (vodcasts using Flash).

Interactive Module

An electronic lesson, often audiovisual, that requires students to complete interactivities.

Case-Based Instruction

The use of patient cases to stimulate discussion, questioning, problem solving, and reasoning on issues pertaining to the basic sciences and clinical disciplines.

Demonstration

A performance or explanation of a process, illustrated by examples, realia, observable action, specimens, and so on.

Discussion or Debate

Instructors facilitate a structured or informal discussion or debate.

Game

An instructional method requiring the learner to participate in a competitive activity with preset rules.

Interview or Panel

Students interview standardized patients or experts to practice interviewing and history-taking skills.

Learning Station

Students rotate through learning stations, participating in performance exercises at each station.

Worksheet or Problem Set

Learners work in pairs or teams to solve problems or categorize information.

CP Scheme

An interactive exercise that encourages learners to make clinical decisions following a clinical presentation scheme (flowchart).

Simulation or Role Play

A method used to replace or amplify real patient encounters with scenarios designed to replicate real health care situations, using lifelike mannequins, physical models, or standardized patients.

Oral Presentation

Students present on topics to their peers. Professors and peers evaluate the presentations using a specific rubric.

Team-Based Activity

A collaborative learning activity that fosters team discussion, thinking, or problem solving.

Problem-Based Learning

Working in peer groups, students identify what they already know, what they need to know, and how and where to access new information that may lead to the resolution of the problem.

Lab or Studio

Students apply knowledge in the lab, by engaging in a hands-on or kinesthetic activity.

Formative Quizzes

The lesson includes a set of questions bundled together into a quiz, which allows learners to self-assess.

Technology-Enhanced Active Learning (TEAL)

An interactive lesson integrating educational technology, such as electronic games, mobile apps, virtual simulations, EHR, videoconferencing, Web exercises, or bioinstruments.

Flipped Classroom

The traditional lecture and homework elements of a course are reversed. Short video lectures or electronic handouts are viewed by students before class. In-class time is devoted to exercises, projects, or discussions. The flipped classroom (also called reverse, inverse, or backwards classroom) is a pedagogical approach in which basic concepts are provided to students for pre-class learning so that class time can be used to apply and build upon those basic concepts. The flipped classroom can be used to prepare students to be lifelong learners and improve their self-reading skills. Furthermore, it improves their basic and clinical knowledge and skills.

Interactive Lecture-Based Teaching Strategy

Pharmacy educators can make the lecture interactive in many ways such as ask students questions every 5 to 10 minutes, present short videos/audios, rally, and team share groups, all of which will attract students to the lectures. Link the theory part with life, share your practice experience with students with mini and long cases and give students time to think about it and solve it. Engage all students and remember that many students may be hesitant to participate, so encourage all to participate. Remember that as the pharmacy educator you are teaching the students and assessing their needs; understanding can help also. Weekly and monthly feedback from students and colleagues can improve online teaching. Record the lectures to give them to students as well as for yourself and colleagues. Add many active teaching strategies to the lecture such as videos, cases, and others.

Blended Teaching Strategy

Traditionally, blended learning combines online educational materials and opportunities for interaction online with traditional place-based classroom methods.

Video-Based Learning

Short videos can be used as an effective teaching strategy for theory, practicals, and training.

Simulation

Role play and other simulation methods can be used with the help of new technologies as an effective online teaching strategy for theory.

Project-Based Learning

Project-based learning (PBL) is a model that can be used to prepare students for real practice. Project-based learning gives students the opportunity to develop knowledge and skills through engaging projects set around challenges and problems they may face in the real world.

Journal Club

A journal club can be used to critically evaluate recent articles in the academic literature.

Case Studies Discussion

Case studies discussion is very important and an effective teaching strategy. Encourage students to read the given cases individually or as teams and solve them.

Self-Directed Learning

Self-directed learning allows students to improve their skills towards self-learning.

Community Services-Based Learning

Many theory courses can be used with this effective teaching strategy, which allow students to achieve the course learning outcomes while contributing to patients, the public, and society.

Seminars

Seminars are an effective strategy and can be used online to improve students' presentation skills.

31.3 Conclusion

This chapter has discussed the traditional and active teaching strategy types in pharmacy education. Traditional teaching strategies were lecture-based and teacher-based, where the educator delivered the lecture and the students listened. Active teaching strategies have been implemented successfully in many pharmacy programs around the world such as team-based learning (TBL) and other active teaching strategies.

References

Al-Meman, A., Al-Worafi, Y.M. and Saeed, M.S., 2014. Team-based learning as a new learning strategy in pharmacy college, Saudi Arabia: Students' perceptions. *Universal Journal of Pharmacy*, 3(3), pp. 57–65.

McCoy, L., Pettit, R.K., Kellar, C. and Morgan, C., 2018. Tracking active learning in the medical school curriculum: A learning-centered approach. *Journal of Medical Education and Curricular Development*, 5, p. 2382120518765135.

Stewart, D.W., Brown, S.D., Clavier, C.W. and Wyatt, J., 2011. Active-learning processes used in US pharmacy education. *American Journal of Pharmaceutical Education*, 75(4).

32

Team-Based Learning in Pharmacy Education

32.1 What Is Team-Based Learning (TBL)?

Team-Based Learning is an evidence-based collaborative learning teaching strategy designed around units of instruction, known as "modules," that are taught in a three-step cycle: preparation, in-class readiness assurance testing, and application-focused exercise. A class typically includes one module (TBL-collaborative, 2021).

32.2 History of Team-Based Learning (TBL)

The history of Team-Based Learning (TBL) goes back to the 1970s when Larry Michaelsen found that his class size had been tripled from 40 to 120 students. He had been using a case-based Socratic teaching approach that involves facilitating problem-solving discussions. He knew that he had two major challenges; the first was to engage a large class in effective problem-solving, and the second was to give his students a reason to prepare before the class session. He developed an approach that is very close to the structure that TBL classrooms use today. He made sure that students came prepared by using an ingenious approach where students were first tested individually, and then in teams. He realized that students were actively discussing the material, which otherwise would have been covered in a lecture, and he devised the "4 S" framework for classroom activities where students worked on a Significant Problem, the Same Problem, where they had to make a Specific Choice, and make a Simultaneous Report. Michaelsen found that this structured problem-solving method for in-class activities really helped to deeply engage students with the content and readily understand how to apply their learning (TBL-collaborative, 2021). In the 1990s TBL became widely recognized and exploited in business schools and many other disciplines in the USA (Michaelsen and Sweet, 2012; Michaelsen, 2002; Al-Meman et al., 2014). Several years later, TBL strategy has been widely employed in medical, nursing, veterinary, dentistry, and health education (Parmelee and Michaelsen, 2010). TBL was successfully implemented in the Ohio State University College of Pharmacy in a workshop teaching a pathophysiology and therapeutics course to encourage better integration of course contents and to endorse regular problem-solving teaching and evaluation (Poirier et al., 2009). A modified TBL format was successfully integrated into a lecture-based cardiovascular module at the school of pharmacy, University of Oklahoma (Conway et al., 2010). The College of Pharmacy, Qassim University, Saudi Arabia implemented TBL for all pharmacotherapy modules, pharmacokinetic, pharmacy profession, introduction to pharmacy practice, and patient assessment courses since 2011 (Al-Meman et al., 2014).

32.3 Importance of Team-Based Learning (TBL)

Team-Based Learning (TBL) has many benefits such as the following (Al-Meman et al., 2014; TBL-collaborative, 2021; Michaelsen and Sweet, 2012):

Improve students' self-learning skills.

Improve students' understanding of the course materials.

Improve students' critical thinking skills.

Improve students' communication skills.

Improve students' attitude towards teamwork and collaboration.

Improve students' decision-making skills.

Improve students' time management skills.

Improve students' presentation skills.

Improve students' problem-solving skills.

Improve knowledge retention.

Improve leadership skills.

Improve ability to answer the cases.

Improve ability to provide/design pharmacist care and patient care.

Improve ability to identify and solve/prevent drug-related problems (DRPs).

32.4 Purpose of Team-Based Learning (TBL)

The aim of TBL is to prepare students to be lifelong learners and to improve their self-reading skills. Furthermore, TBL improves their basic and clinical knowledge and skills.

32.5 Elements of Team-Based Learning (TBL)

The four essential elements of TBL are as follows (Sibley and Ostafichuk, 2015):

1. Teams must be properly formed and managed.
2. Students must be motivated to come to class prepared.
3. Students must learn to use course concepts to solve problems.
4. Students must be truly accountable.

32.6 Principles of Team-Based Learning (TBL)

32.6.1 Principle 1—Groups Must Be Properly Formed and Managed

Groups need to be formed in a way that enables them to do the work that they will be asked to do.

Minimize barriers to group cohesiveness.

Distribute member resources. In order to function as effectively as possible, each group should have access to whatever assets exist within the whole class and not carry more than a "fair share" of the liabilities.

Learning teams should be fairly large and diverse; the teams should comprise five to seven members.

Groups should be permanent; groups should be equitable based on the academic and linguistic abilities of their members.

32.6.2 Principle 2—Students Must Be Made Accountable

Developing groups into cohesive learning teams requires assessing and rewarding a number of different kinds of student behavior. Students must be accountable for (a) individually preparing for group work; (b) devoting time and effort to completing group assignments; and (c) interacting with each other in productive ways. Fortunately, team learning offers opportunities to establish each of these three forms of accountability:

Accountability for individual pre-class preparation.

Accountability for contributing to the team.

Accountability for high-quality team performance.

Grading System: It is essential that you use an overall assessment system for the course that encourages the kind of student behavior that will promote learning in and from group interaction.

32.6.3 Principle 3—Team Assignments Must Promote Both Learning and Team Development

The development of appropriate group assignments is a critical aspect of successfully implementing team learning.

32.6.4 Principle 4—Students Must Receive Frequent and Immediate Feedback

For teams to perform effectively and to develop as a team, they must have regular and timely feedback on group performance.

Timely feedback from the Readiness Assessment Tests (RATs) is an important source of feedback that supports both learning and team development.

Timely feedback on application-focused team assignments, for example, providing immediate feedback on application-focused team assignments is also important for both learning and team development.

32.7 Tips for Implement Effective Team-Based Learning (TBL)

1. Start with good course design.
2. Use a "backward design" when developing a TBL course and modules.
3. Make sure you organize the module activities so that students can reach your learning goals and you (and they) will know that they have done it.
4. Have application exercises that promote both deep thinking and engaged content-focused discussion.
5. Do not underestimate the importance of the RAP (readiness assurance process).
6. Orient the class as to why you are using TBL and how it is different from previous experiences they may have had with learning groups.
7. Highlight accountability as the cornerstone of TBL.
8. Provide a fair appeals process that will inspire further learning.
9. Peer evaluation is a challenge to get going, but it can enhance the accountability of the process.
10. Be clear and focused with the advanced preparation.
11. Create the team thoughtfully.
12. Several low-budget "props" can facilitate the implementation of a good module.

32.8 Tips for Designing Effective Team-Based Learning (TBL) for Any Course

Orientation is very important for the success of TBL. Orientation should be about the course objectives, learning outcomes, topics, educational resources, and the TBL process.

Provide the students with the course resources such as books, lecture notes, and others.

Make a mock TBL and get feedback from all students and peers.

Design the TBL session based on the course credit hours, allocated time for the course per week, and adjust accordingly.

A TBL session usually consists of the following:

- IRAT (individual readiness assurance test): in this step, the students will take the exam individually and solve it in 15 minutes as an example; you may adjust it based on the course and need. The type of questions could be multiple-choice questions (MCQs) as this is reported, or in my opinion, the lecturer may change it to any type such as short essay or any form. The most important issue is to achieve the course learning outcome and to improve students' knowledge and skills. However, 5 to 15 MCQs could be an option for 10 to 15 minutes.

- gRAT/tRAT (group or team readiness assurance test): in this step the same iRAT questions are redistributed to all groups for discussion, answering, and justifying the answers. The time allowed for gRAT is 30 minutes. The students are encouraged to vote for the best correct answer during the gRAT, and if there is no agreement, to enhance discussion and interaction among the group members.

- The next 20 minutes are considered for the in-place presentation of the answers. The presentation of the answer must be done simultaneously among the groups (at the same time) by voting to make it more interactive, followed by justifying and defending against the answers.

- The next 20 minutes are allocated for open discussion.

- The last 10 minutes are assigned for peer evaluation and appeals. Encourage students to evaluate each teammate on their contributions to the team's success and their own learning. Developing an evaluation checklist with both qualitative and quantitative questions could be easier and facilitate the TBL session.

- Break for 20 minutes, if the lecture is more than 2 hours, otherwise, the last phase will be conducted on another day.

- Application activities for 2 hours: in this step, the predeveloped cases or exercises will be distributed to the students as groups and students will be asked to answer them as teams.
- The grading system for the TBL session allocates marks for each step. Example: 10 marks for iRAT, 15 marks for gRAT, 5 marks for the peer evaluation, 20 marks for the cases, or you may adjust it as needed.

32.9 Tips for Implementing Online Team-Based Learning (TBL)

Team-based learning (TBL) can be implemented online, or as a hybrid (face to face with online). The following steps are very important in implementing TBL online:

Check the technology facilities at your school as well as at students' homes.

Training about the technology use for online TBL is very important to pharmacy educators as well as students.

Orientation is very important for the success of TBL. Orientation should be about the course objectives, learning outcomes, topics, educational resources, and the TBL process.

Provide the students with online course resources such as e-books, lecture notes, and others.

Make a mock online TBL and get feedback from all students and peers.

Design the TBL session based on the course credit hours and the allocated time for the course per week and adjust it accordingly.

A TBL session usually consists of the following:

- Online IRAT (individual readiness assurance test) by using Blackboard or Moodle as well as Microsoft Teams or any video conferencing platform to monitor the iRAT. For this step you may make one Microsoft Teams link for all students. In this step, the students will take the exam individually and solve it in 15 minutes as an example; you may adjust it based on the course and need. The type of questions could be multiple- choice questions (MCQs) as this is reported or, in my opinion, the lecturer may change it to any type such as short essay or any form. The most important issue is to achieve the course learning outcome and to improve students' knowledge and skills. However, 5 to 15 MCQs could be an option for 10 to 15 minutes.
- Online gRAT/tRAT (group or team readiness assurance test) by using Blackboard or Moodle as well as Microsoft Teams or any video conferencing platform to monitor the gRAT. At this step you may

make one Microsoft Teams link for all students and distribute the teams through Microsoft Teams. In this step, the same iRAT questions are redistributed to all groups for discussion, answering, and justifying the answers. The time allowed for gRAT is 30 minutes. Students are encouraged to vote for the best correct answer during the gRAT. If there is no agreement, they are encouraged to enhance discussion and interaction among the group members.

- The next 20 minutes are considered for the in-place presentation of the answers. At this step you may make one Microsoft Teams link for all students. The presentation of the answer must be done simultaneously among the groups (at the same time) by voting to make it more interactive, followed by justifying and defending against the answers.
- The next 20 minutes are allocated for open discussion.
- The last 10 minutes are assigned for peer evaluation and appeals. Encourage students to evaluate each teammate on their contributions to the team's success and their own learning. Developing an evaluation checklist with both qualitative and quantitative questions could be easier and facilitate the TBL session.
- Break for 20 minutes, if the lecture is more than 2 hours. Otherwise, the last phase will be conducted on another day.
- Application activities for 2 hours. At this step you may make one Microsoft Teams link for all students. In this step, the predeveloped cases or exercises will be distributed to the students as groups and students will be asked to answer them as teams.
- The grading system for the TBL session allocates marks for each step. Example: 10 marks for iRAT, 15 marks for gRAT, 5 marks for the peer evaluation, 20 marks for the cases, or you may adjust as needed.

32.10 Barriers to Implementing Online Team-Based Learning (TBL)

There are many barriers to implementing online team-based learning (TBL) as follows:

Knowledge of pharmacy educators about TBL.

Attitude of pharmacy educators about TBL.

Attitude of pharmacy students about TBL.

Resistance of pharmacy educators to TBL.

Resistance of pharmacy students to TBL.

University/school culture towards new teaching strategies.

Lack of training about TBL.

Lack of motivation to train about TBL.

Lack of technologies to implement online TBL.

32.11 Conclusion

This chapter has discussed the history and importance of Team-Based Learning (TBL); purpose of TBL; Process of TBL; Tips for implement an effective Team-Based Learning (TBL); Tips for implement an effective online Team-Based Learning (TBL); and barriers for conducting Team-Based Learning (TBL).

References

Al-Meman, A., Al-Worafi, Y.M. and Saeed, M.S., 2014. Team-based learning as a new learning strategy in pharmacy college, Saudi Arabia: Students' perceptions. *Universal Journal of Pharmacy*, 3(3), pp. 57–65.

Conway, S.E., Johnson, J.L. and Ripley, T.L., 2010. Integration of team-based learning strategies into a cardiovascular module. *American Journal of Pharmaceutical Education*, 74(2).

Michaelsen, L.K., 2002. *Getting started with team-based learning. Team-based learning: A transformative use of small groups*, pp. 27–51. Available at: https://iucat.iu.edu/iuk/6433622

Michaelsen, L.K. and Sweet, M., 2012. Fundamental principles and practices of team-based learning. In *Team-based learning for health professions education: A guide to using small groups for improving learning* (pp. 9–34).

Parmelee, D.X., DeStephen, D. and Borges, N.J., 2009. Medical students' attitudes about team-based learning in a pre-clinical curriculum. *Medical Education Online*, 14(1), p. 4503.

Parmelee, D.X. and Michaelsen, L.K., 2010. Twelve tips for doing effective team-based learning (TBL). *Medical Teacher*, 32(2), pp. 118–122.

Poirier, T.I., Butler, L.M., Devraj, R., Gupchup, G.V., Santanello, C. and Lynch, J.C., 2009. A cultural competency course for pharmacy students. *American Journal of Pharmaceutical Education*, 73(5).

Sibley, J. and Ostafichuk, P., 2015. *Getting started with team-based learning*. Stylus Publishing, LLC.

Team-Based Learning Collaborative (TBL-collaborative), 2021. Available at: www.teambasedlearning.org/

33

Problem-Based Learning in Pharmacy Education

33.1 What Is Problem-Based Learning (PBL)?

Problem-based learning (PBL) is a pedagogical approach that enables students to use "triggers" from the problem case or scenario to define their own learning objectives. Subsequently they do independent, self-directed study before returning to the group to discuss and refine their acquired knowledge. Thus, PBL is not about problem solving per se, but rather it uses appropriate problems to increase knowledge and understanding. The process is clearly defined, and the several variations that exist all follow a similar series of steps (Wood, 2003). Barrows (1996) stated the following characteristics of problem-based learning (Barrows, 1996):

Learning is student-centered under the guidance of a tutor.

Learning occurs in small student groups. In most of the early PBL medical schools, groups were made up of five to eight or nine students. Characteristically, at the end of each curricular unit, the students are resorted randomly into new groups with a new tutor.

Teachers are facilitators or guides.

Problems form the organizing focus and stimulus for learning.

Problems are a vehicle for the development of clinical problem-solving skills.

New information is acquired through self-directed learning.

33.2 History of Problem-Based Learning (PBL)

The history of problem-based learning (PBL) goes back to the 1960s when the PBL process was pioneered by Barrows and Tamblyn at the medical

DOI: 10.1201/9781003230458-37

school program at McMaster University in Hamilton (Barrows, 1996). Several curricular innovations in pharmacy schools have utilized PBL methods to teach pharmacotherapy, pharmacokinetics, and others since 1990 (Culbertson et al., 1997).

33.3 Importance of Problem-Based Learning (PBL)

Problem-based learning (TBL) has many benefits such as the following (Barrows, 1996; Wood, 2003):

The acquisition of an integrated knowledge base.

The acquisition of a knowledge base structured around the cues presented by patient problems.

The acquisition of a knowledge base enmeshed with problem-solving processes used in clinical medicine.

The development of an effective and efficient clinical problem-solving process.

The development of effective self-directed learning skills. The development of team skills.

Improved student understanding of the course materials.

Improved students leadership (chairing a group).

Improved students listening skills.

Improved students recording skills.

Improved students communication skills.

Improved student attitudes towards teamwork and collaboration.

Improved student "respect for colleagues" views.

Improved student critical evaluation of literature.

Improved student use of resources.

Improved student presentation skills.

33.4 Purpose of Problem-Based Learning (PBL)

The aim of problem-based learning is to prepare the students to be lifelong learners and to improve their self-reading skills. Furthermore, students will improve their basic and clinical knowledge and skills.

33.5 The Problem-Based Learning (PBL) Process

Wood (2003) reported the following steps:

Step 1—Identify and clarify unfamiliar terms presented in the scenario; the scribe lists those that remain unexplained after discussion.

Step 2—Define the problem or problems to be discussed; students may have different views on the issues, but all should be considered; the scribe records a list of agreed problems.

Step 3—"Brainstorming" session to discuss the problem(s), suggesting possible explanations on the basis of prior knowledge; students draw on each other's knowledge and identify areas of incomplete knowledge; the scribe records all discussion.

Step 4—Review steps 2 and 3 and arrange explanations into tentative solutions; the scribe organizes the explanations and restructures if necessary.

Step 5—Formulate learning objectives; the group reaches consensus on the learning objectives; the tutor ensures that the learning objectives are focused, achievable, comprehensive, and appropriate.

Step 6—Private study (all students gather information related to each learning objective).

Step 7—Group shares results of private study (students identify their learning resources and share their results); the tutor checks learning and may assess the group.

Trigger material for PBL scenarios are as follows:

Paper-based clinical scenarios

Experimental or clinical laboratory data

Photographs

Video clips

Newspaper articles

All or part of an article from a scientific journal

A real or simulated patient

A family tree showing an inherited disorder

Others

33.6 Tips for Designing Effective Problem-Based Learning (PBL)

Wood (2003) reported the following tips for designing effective PBL:

Learning objectives likely to be defined by the students after studying the scenario should be consistent with the faculty learning objectives.

Problems should be appropriate to the stage of the curriculum and the level of the students' understanding.

Scenarios should have sufficient intrinsic interest for the students or relevance to future practice.

Basic science should be presented in the context of a clinical scenario to encourage integration of knowledge.

Scenarios should contain cues to stimulate discussion and encourage students to seek explanations for the issues presented.

The problem should be sufficiently open so that discussion is not curtailed too early in the process.

Scenarios should promote participation by the students in seeking information from various learning resources.

33.7 Disadvantages of Problem-Based Learning (PBL)

Wood (2003) reported the following disadvantages of PBL:

Tutors who can't "teach"—Tutors

enjoy passing on their own knowledge and understanding and so may find PBL facilitation difficult and frustrating.

Human resources—More staff

have to take part in the tutoring process.

Other resources—Large numbers

of students need access to the same library and computer resources simultaneously.

Role models—Students may be

deprived of access to a particular inspirational teacher who in a traditional curriculum would deliver lectures to a large group.

Information overload—Students

may be unsure how much self-directed study to do and what information is relevant and useful.

33.8 Tips for Implementing Online Problem-Based Learning (PBL)

Problem-based learning (PBL) can be implemented online, or as a hybrid (face to face with online). The following steps are very important to implementing TBL online:

Check the technology facilities at your school as well as at students' homes.

Training about technology use for online PBL is very important to pharmacy educators as well as students.

Orientation is very important for the success of PBL.

Follow the above steps for PBL.

33.9 Barriers to Implementing Problem-Based Learning (PBL)

There are many barriers for implementing problem-based learning (PBL) as follows:

Lack of resources.

Knowledge of pharmacy educators about PBL.

Attitude of pharmacy educators about PBL.

Attitude of pharmacy students about PBL.

Resistance of pharmacy educators about PBL.

Resistance of pharmacy students about PBL.

University/school culture towards the new teaching strategies.

Lack of training about PBL.

Lack of motivation to train about PBL.

Lack of technologies to implement online PBL.

33.10 Conclusion

This chapter has discussed the history and importance of problem-based learning (PBL); the purpose of PBL; elements of PBL; the process of PBL; tips for implementing effective PBL; Tips for implementing effective online PBL; and barriers to conducting PBL.

References

Barrows, H.S., 1996. Problem-based learning in medicine and beyond: A brief overview. *New Directions for Teaching and Learning*, 1996(68), pp. 3–12.

Culbertson, V.L., Kale, M. and Jarvi, E.J., 1997. Problem-based learning: A tutorial model incorporating pharmaceutical diagnosis. *American Journal of Pharmaceutical Education*, 61(1), pp. 18–25.

Wood, D.F., 2003. Problem based learning. *BMJ*, 326(7384), pp. 328–330.

34

Case-Based Learning in Pharmacy Education

34.1 What Is Case-Based Learning (CBL)?

Case-based learning (CBL) is a pedagogical approach using real or simulation case studies, allowing students to apply their knowledge and critical thinking to answer the cases. A case-based learning approach will prepare students for real clinical, pharmaceutical care/pharmacist and patient care practice. Cases can be related to the majority of pharmacy courses such as pharmacotherapy modules, clinical pharmacokinetics, basic pharmacokinetics, pharmacy practice, pharmaceutical care, pharmacology, ethics and law, and other courses.

34.2 History of Case-Based Learning (CBL)

The history of case-based learning (CBL) in pharmacy education goes back to the 1990s (Reddy, 2000).

34.3 Importance of Case-Based Learning (CBL)

Case-based learning (CBL) has many benefits for students as follows:

Prepares students for real practice.
Improves students' critical thinking skills.
Improves students' communication skills.
Improves students' presentation skills.
Improves students' reading skills.
Improves students' searching/literature review skills.

DOI: 10.1201/9781003230458-38

Improves students' critical evaluation of literature.

Improves students' ability to design the appropriate management plan for patients.

Improves students' ability to practice evidence-based medicine/practice.

34.4 Purpose of Case-Based Learning (CBL)

The aim of case-based learning is to prepare students for real-life practice; furthermore, to improve their competencies towards providing the most effective patient care.

34.5 Tips for Designing Effective Case-Based Learning (CBL)

Prepare the learning outcomes for the cases. This could be different from one course to another.

Design or adapt the cases from hospitals, real life (real cases), or from course references.

Distribute the students to groups, each group with five to seven students.

Give the cases to students and explain it to them, and what they should do.

Allocate time to answer the cases.

Encourage students to discuss the answers and defend their answers.

Provide feedback to students to improve their knowledge and skills.

34.6 Tips for Implementing Online Case-Based Learning (CBL)

Case-based learning (CBL) can be implemented online, or as a hybrid (face to face with online). The following steps are very important to implementing case-based learning (CBL) online:

Check the technology facilities at your school as well as at students' homes.

Training about technology use for online CBL is very important to pharmacy educators as well as students.

Orientation is very important for the success of CBL.

Follow the above steps for CBL.

35.7 Barriers to Implementing Case-Based Learning (CBL)

There are many barriers to implementing case-based learning (CBL) as follows:

Knowledge of pharmacy educators about CBL.

Attitude of pharmacy educators about CBL.

Attitude of pharmacy students about CBL.

Resistance of pharmacy educators to CBL.

Resistance of pharmacy students to CBL.

University/school culture towards new teaching strategies.

Lack of training about CBL.

Lack of motivation to train about CBL.

Lack of technologies to implement online CBL.

34.8 Forms of Case-Based Learning (CBL)

Case-based learning can be in short case form or long case form. Cases are different from one course to another, and it depends on the course learning outcomes, allocated time for the tutorial or case discussion, and other factors. In short cases, you present the case with patient's related information, assessment, and diagnosis and ask students to design the management plan for the patient. Note: The management plan will include goals of therapy and desired outcomes for all diseases/conditions; non-pharmacological therapies (individualize the "lifestyle changes" such as weight control, healthy dietary therapy, increased physical activity, modifying the modifiable risk factors, etc. depending on the disease and patient situation); pharmacological therapies (appropriate and rational based on the guidelines) recommendations with doses, dosage form and route of administration, strength, frequency, duration; time of taking medications and instructions; monitoring parameters: the efficacy of medications (is the prescribed medication effective; is the desired outcome achieved). This can be done by using the laboratory results, checking the symptoms' improvement, patients' report and other criteria; the safety of medications (is the prescribed medication safe). This can be done by patients' reports about side/adverse effects/reactions, evaluating the effects patient's different systems such as renal, liver, etc., requesting laboratory tests, requesting drug therapy monitoring (TDM), and others.

Adherence towards the management plan; therapy success and complications: Is the treatment desired outcome achieved?; patient education and counseling related to adherence towards the management plan (non-pharmacological, pharmacological therapies, and monitoring parameters), self-management, potential adverse drug effects and reactions, possible interactions, cautions and precautions, contraindications and warnings, proper storage and disposal of medications.

Long Case Example in Pharmacotherapy as Follows:

Case of XXXXXXXX (Any Disease)

Patient data: Age: years, weight kg, height: cm, gender: male/female.

Chief Complaint:

History of present illness:

Past Medical History:

Medications history:

Medication name	Dose	Dosage form	Frequency/ duration	Indication

Family history:

Social history:

Allergies:

Review of systems:

Blood pressure	Day 1	Day 2
Heart rate		
Respiratory rate		
Temperature		
Weight		
Height		
BMI		
CrCl		

Physical Examination

General appearance:

Skin:

HEENT:

Neck:

Lungs:

CV:

Abd:

Genit/Rect:

MS/Ext:

Neuro:

ECG:

Labs:

Hematology:

Note: Please check the reference (normal) ranges from the hospital, as each hospital could use another reference.

Lab	Normal range	Day 1	Day 2	Day 3	Day 4	Day 5	Day 6
WBC							
RBC							
HGB							
HCT							
MCV							
MCH							
MCHC							
RDW-CV							
PLT							
NE %							
LY%							
MO %							

Lab	Normal range	Day 1	Day 2	Day 3	Day 4	Day 5	Day 6
EO %							
BA %							
NE #							
LY #							
MO #							
EO #							
BA #							
EO #							
BA #							

Renal profile

Lab	Normal range	Day 1	Day 2	Day 3	Day 4	Day 5	Day 6
SODIUM							
POTASSIUM							
UREA							
Sr Cr							
Cl Cr							

Liver function tests

Lab	Normal range	Day 1	Day 2	Day 3	Day 4	Day 5	Day 6
TOTAL PROTEIN							
ALBUMIN							
GLOBULIN							
A/G RATIO							
TOTAL BILIRUBIN							
ALT							
ALP							

Cardiac enzymes tests

Lab	Normal range	Day 1	Day 2	Day 3	Day 4	Day 5	Day 6
ASPARTAT TRANSAMINASE							
CREATINE KINASE							
LACTAT DEHYDROGENASE							

Blood glucose monitoring chart

	Day 1/ time	Day1/ time	Day2/ time	Day2/ time		
HBA1C						
RBG						
FBG						

Lipid profile

		Day 1						
TC								
LDL								
HDL TG								

Assessment
Daily follow-up

Day 1 SOAP
    Subjective
    Objective
    Assessment
    Plan

Day 2 SOAP
    Subjective
    Objective
    Assessment
    Plan

Day 3 SOAP
    Subjective
    Objective
    Assessment
    Plan

Drug-Related Problems (DRPs)
DRP 1
Justification

Recommendation

DRP 2
Justification

Recommendation

DRP 3
Justification

Recommendation

34.9 Conclusion

This chapter has discussed the history and importance of case-based learning (CBL); purpose of CBL; process of CBL; tips for implementing effective CBL; tips for implementing effective online CBL; and barriers to conducting CBL.

Reference

Reddy, I.K., 2000. Implementation of a pharmaceutics course in a large class through active learning using quick-thinks and case-based learning. *American Journal of Pharmaceutical Education*, 64(4), pp. 348–354.

35

Simulation in Pharmacy Education

35.1 What Is Simulation and Simulation-Based Education?

Medical simulation, or more broadly, health care simulation, is a branch of simulation related to education and training in medical fields of various industries. Simulations can be held in the classroom, in situational environments, or in spaces built specifically for simulation practice. It can involve simulated human patients—artificial, human, or a combination of the two; educational documents with detailed simulated animations; casualty assessment in homeland security and military situations; emergency response; and support virtual health functions with holographic simulation. In the past, its main purpose was to train medical professionals to reduce errors during surgery, prescription, crisis interventions, and general practice. Combined with methods in debriefing, it is now also used to train students in anatomy, physiology, and communication during their schooling (Gaba, 2004a, b; Fanning and Gaba, 2007). The term virtual reality was introduced to describe immersive environments. "Simulator" refers to a physical object or representation of the full or part task to be replicated. "Simulation" refers to applications of simulators for education or training. The term simulator is used by some specifically to refer to technologies that recreate the full environment in which one or more targeted tasks are carried out. This can also be called fully immersive simulation. The term "part-task trainer" should be applied to technologies that replicate only a portion of a complete process or system. However, simulator is commonly used in a generic sense to apply to all technologies that are used to imitate tasks (Cooper and Taqueti, 2008). The Accreditation Council for Pharmacy Education (ACPE, 2018) defined simulation as an activity or event replicating pharmacy practice. For the purpose of satisfying Introductory Pharmacy Practice Experiences (IPPE) expectations, simulation includes multiple types of scenarios based on the utilization of high-fidelity manikins, medium-fidelity manikins, standardized patients, role playing, Objective Structured Clinical Evaluations (OSCE), and computer-based simulations. Simulation as a component of IPPE should clearly connect the pharmacy activity or delivery of a medication to a patient (whether simulated patient, standardized patient, or virtual patient) (ACPE, 2018).

DOI: 10.1201/9781003230458-39

35.2 History of Simulation

The history of simulation-based learning (SBL) in medical education goes back to the 18th century Paris, when Grégoire father and son developed an obstetrical mannequin made of a human pelvis and a dead baby. The phantom, as the mannequin was named, enabled obstetricians to teach delivery techniques, which resulted in a reduction of maternal and infant mortality rates (Rosen, 2008). The process of using patient actors to educate began in 1963 (Rosen, 2008); the Laerdal product was one of the first significant events in the history of medical simulation. She was initially designed for the practice of mouth-to-mouth breathing. Her face was based on the death mask of the Girl from the River Seine, a famous French drowning victim. Laerdal wanted to encourage the practice of rescue techniques by the design of a sympathetic simulated victim. Annie evolved to incorporate a spring in her chest for the practice of cardiopulmonary resuscitation (CPR). The cardiology patient simulator, Harvey, debuted at the University of Miami a few years later. Simple plastic models for the practice of resuscitation, physical examination, and procedural skills appeared over the next few decades. The next significant advances did not occur until the 1990s (Rosen, 2008). The history of simulation in pharmacy education goes back to the 1990s (Rao, 2011). Role-play has been used primarily as a means of helping students develop skills in communication, consultation, and medication history-taking; and as a tool for assessing the effectiveness of training programs (Rao, 2011). The Accreditation Council for Pharmacy Education approved simulation for the purpose of satisfying Introductory Pharmacy Practice Experiences (IPPE) expectations. Simulation includes multiple types of scenarios based on the utilization of high-fidelity manikins, medium-fidelity manikins, standardized patients, role playing, Objective Structured Clinical Evaluations (OSCE), and computer-based simulations. Simulation as a component of IPPE should clearly connect the pharmacy activity or delivery of a medication to a patient (whether simulated patient, standardized patient, or virtual patient) (ACPE, 2018).

35.3 Importance of Simulation

Simulation-based learning (SBL) has many benefits to students as follows (ACPE, 2018; Lateef, 2010; Al-Worafi, 2020; Rao, 2011; Coyne et al., 2019):

> Simulation-based learning can be the way to develop health professionals' knowledge, skills, and attitudes, while protecting patients from unnecessary risks.

Simulation-based medical education can be a platform which provides a valuable tool in learning to mitigate ethical tensions and resolve practical dilemmas.

Prepares students for real practice.

Improves students' communication skills.

Improves students' ability to take patient history.

Improves students' ability to perform/document patient assessment.

Improves students' ability to identify patient needs.

Improves students' ability to identify drug-related problems (DRPs).

Improves students' ability to design a plan to manage actual DRPs.

Improves students' ability to design a plan to prevent/minimize potential DRPs.

Improves students' ability to design a management plan for patients, which includes goals of therapy and desired outcomes for all diseases/conditions.

Non-pharmacological therapies (individualize "lifestyle changes" such as weight control, healthy dietary therapy, increased physical activity, modifying the modifiable risk factors, etc. depending on the disease and patient situation). Pharmacological therapies (appropriate and rational based on the guidelines) recommendations with doses, strength, dosage form and route of administration, frequency, duration; time of taking medications and instructions).

Appropriate monitoring parameters: The efficacy of medications (is the dispensed medication effective; is the desired outcome achieved). This can be done by using/recommending laboratory results, checking the symptoms' improvement, patients' report, and other criteria. The safety of medications (is the dispensed medication safe). This can be done by patients' reports about side/adverse effects/reactions, evaluating the effects on patients' different systems such as renal, liver, etc., requesting/recommending laboratory tests, requesting drug therapy monitoring (TDM), and others. Adherence towards the management plan.

Therapy success and complications: Is the treatment desired outcome achieved?

Appropriate patient education and counseling.

Improves students' ability to conduct effective patient education and counseling, which includes:

Educate and counsel patients/the public about the appropriate use of self-medications.

Identify patients/the public at risk for medications misuse and abuse, and educate and counsel them about the effects of medications misuse and abuse on their health.

Educate and counsel patients/the public about the appropriate way to store their medications.

Educate and counsel patients/the public about the appropriate way to dispose of expired and unused medications.

Improves students' ability towards medication safety practice issues such as reporting adverse drug reactions (ADRs); reporting medication errors and other safety practices.

Improves students' critical thinking skills.

Improves students' presentation skills.

Improves students' reading skills.

Improves students' searching/literature review skills.

Improves students' critical evaluation of literature.

Improves students' ability to practice evidence-based medicine/practice.

Improves students' ability to dispense medications.

Provides immediate feedback from educators.

Improves students' ability to assess pharmacy practice issues.

35.4 Purpose of Simulation-Based Learning (SBL)

The aim of simulation-based learning is to prepare students for real-life practice; furthermore, to improve their competencies towards providing the most effective patient care.

35.5 Types of Simulation

There are many types of simulation as follows:

- Manikin-based simulation.
- Skills-training simulation.
- Virtual reality simulation.
- Standardized patient simulation.
- Software simulation.

35.6 Applications of Simulation

Gaba (2004b) reported that simulation applications can be categorized by 11 dimensions as follows (Gaba, 2004b):

Dimension 1: The purpose and aims of the simulation activity.

The most obvious application of simulation is to improve the education and training of clinicians, but other purposes are also meaningful. *Education* emphasizes conceptual knowledge, basic skills, and an introduction to the actual work. *Training* emphasizes the actual tasks and work to be performed.

Dimension 2: The unit of participation in the simulation.

Many simulation applications are targeted at individuals. These may be especially useful for teaching knowledge and basic skills.

Dimension 3: The experience level of simulation participants.

Simulation can be applied from "cradle to grave" of clinical personnel.

Dimension 4: The health care domain in which the simulation is applied.

Simulation techniques can be applied across nearly all health care domains.

Dimension 5: The health care disciplines of personnel participating in the simulation.

Simulation is applicable to all disciplines of health care.

Dimension 6: The type of knowledge, skill, attitudes, or behavior addressed in simulation.

Simulations can be used to help learners acquire new knowledge, and to better understand conceptual relations and dynamics.

Dimension 7: The age of the patient being simulated.

To date, the bulk of simulators and simulation applications have been addressed to adult patients and clinical activities relevant to adult medicine. Simulation may be particularly useful for pediatric patients and clinical activities.

Dimension 8: The technology applicable or required for simulations.

To accomplish these goals a variety of technologies (including no technology) will be relevant for simulation. For example, some education and training on teamwork can be accomplished with role playing or analysis of videos.

Dimension 9: The site of simulation participation.

Some types of simulation—those that use videos, computer programs, or the Web—can be conducted in the privacy of the learner's home or office using their own personal computer. More advanced screen-based simulators might need more powerful computer facilities available in the medical library.

Dimension 10: The extent of direct participation in simulation.

Most simulations—even screen-based simulators or part-task trainers—were initially envisioned as highly interactive activities with significant direct "on site" hands-on participation. However, not all learning requires direct participation. For example, some learning can take place merely by viewing a simulation involving others.

Dimension 11: The feedback method accompanying simulation.

Much as in real life, one can learn a great deal just from the experience itself, without any additional feedback. For most complex simulations, specific feedback is provided to maximize learning. On screen-based simulators or virtual reality systems, the simulator itself can provide feedback about the participant's actions or decisions.

35.7 Tips for Designing Effective Simulation-Based Learning (SBL)

Prepare the learning outcomes for the simulation. This could be different from one course to another.

Select the type of simulation such as role play or other types.

Training about the SimMan and other simulation types is very important for educators to succeed in simulation-based learning.

Design or adapt the simulation scenarios.

Orientation is very important.

Explain the theory part to students if needed or ask them to read about it before the simulation session.

Encourage students to participate, observe, and practice.

Provide feedback to students to improve their knowledge and skills.

Recording the session with permission can help in providing feedback.

35.8 Tips for Implementing Online Simulation-Based Learning (SBL)

Simulation-based learning can be implemented online as role play or software simulation.

The following steps are very important to implement simulation-based learning online:

Check the technology facilities at your school, educators' homes and students' homes.

Training about technology use for online SBL is very important for pharmacy educators as well as students.

Prepare the learning outcomes for the simulation; this could be different from one course to another.

Select the type of simulation such as role play or other types.

Training about the SimMan and other simulation types is very important for educators to succeed in simulation-based learning.

Design or adapt the simulation scenarios.

Orientation is very important.

Explain the theory part to students if needed or ask them to read about it before the simulation session.

Encourage students to participate, observe, and practice.

Provide feedback to students to improve their knowledge and skills.

Recording the session with permission can help in providing feedback.

35.9 Barriers to Implementing Simulation-Based Learning (SBL)

There are many barriers to implementing simulation-based learning (SBL) as follows:

Lack of resources.

Lack of funds, financial issues.

Knowledge of pharmacy educators about SBL.

Attitude of pharmacy educators about SBL.

Attitude of pharmacy students about SBL.

Resistance of pharmacy educators towards SBL.

Resistance of pharmacy students towards SBL.

University/school culture towards SBL.

Lack of training about SBL.

Lack of motivation to train about SBL.

Lack of technologies to implement online SBL.

35.10 Conclusion

This chapter has discussed the history and importance of simulation-based learning (SBL); the purpose of SBL; the process of SBL; tips for implementing effective SBL; tips for implementing effective online SBL; and barriers to conducting SBL.

References

Accreditation Council for Pharmacy Education (ACPE), 2018. *Policies and procedures for ACPE accreditation of professional degree programs.* Available at: https://www.acpe-accredit.org/pdf/PoliciesandProcedures.pdf

Al-Worafi, Y.M., 2020. Quality indicators for medications safety. In *Drug safety in developing countries* (pp. 229–242). Academic Press.

Cooper, J.B. and Taqueti, V., 2008. A brief history of the development of mannequin simulators for clinical education and training. *Postgraduate Medical Journal*, 84(997), pp. 563–570.

Coyne, L., Merritt, T.A., Parmentier, B.L., Sharpton, R.A. and Takemoto, J.K., 2019. The past, present, and future of virtual reality in pharmacy education. *American Journal of Pharmaceutical Education*, 83(3).

Fanning, R.M. and Gaba, D.M., 2007. The role of debriefing in simulation-based learning. *Simulation in Healthcare*, 2(2), pp. 115–125.

Gaba, D.M., 2004a. A brief history of mannequin-based simulation and application. *Simulators in Critical Care and Beyond*, pp. 7–14.

Gaba, D.M., 2004b. The future vision of simulation in health care. *BMJ Quality & Safety*, 13(Suppl 1), pp. i2–i10.

Lateef, F., 2010. Simulation-based learning: Just like the real thing. *Journal of Emergencies, Trauma and Shock*, 3(4), p. 348.

Rao, D., 2011. Skills development using role-play in a first-year pharmacy practice course. *American Journal of Pharmaceutical Education*, 75(5).

Rosen, K.R., 2008. The history of medical simulation. *Journal of Critical Care*, 23(2), pp. 157–166.

36

Project-Based Learning in Pharmacy Education

36.1 What Is Project-Based Learning?

Cocco, 2006 defined project-based learning (PBL) as a student-centered form of instruction which is based on three constructivist principles: learning is context-specific, learners are involved actively in the learning process, and they achieve their goals through social interactions and the sharing of knowledge and understanding (Cocco, 20).

36.2 History of Project-Based Learning

The history of project-based learning (PBL) goes back to 1897. Dewey wrote a book in 1897 called *My Pedagogical Creed*, which outlined the concept of "learning by doing." While many teachers embrace Dewey's writings as the true birth of project-based learning, a quick review of history shows things a bit differently (Kokotsaki et al., 2016).

36.3 Importance of Project-Based Learning

Project-based learning (PBL) has many benefits for students as follows (Kokotsaki et al., 2016):

Preparing students for real practice, project-based learning (PBL) gives students the opportunity to develop knowledge and skills through engaging projects set around challenges and problems they may face in the real world.

DOI: 10.1201/9781003230458-40

Improves students' communication skills.

Improves students' critical thinking skills.

Improves students' presentation skills.

Improves students' reading skills.

Improves students' searching/literature review skills.

Improves students' critical evaluation of literature

Improves students' ability to conduct research.

Improves students' self-learning skills.

36.4 Purpose of Project-Based Learning (PBL)

Project-based learning is designed to give students the opportunity to develop knowledge and skills through engaging projects set around challenges and problems they may face in the real world. The goals of project-based self-directed learning are (Keator et al., 2016):

1. Identify gaps in knowledge as related to science and medicine.
2. Identify a learning topic to address gaps in knowledge.
3. Develop learning objectives for the identified learning topic.
4. Retrieve relevant information resources.
5. Evaluate information sources.
6. Use evidence-based medicine to distinguish facts from fiction.
7. Collaborate with peers to create a deliverable.
8. Provide constructive feedback to peers.
9. Receive critical feedback from a diverse audience.
10. Practice reflection to independently identify learning preferences, strengths, weaknesses, and personal bias.

36.5 Elements of Project-Based Learning (PBL)

The elements of project-based learning (PBL) are (Ms, 2010):

A Need to Know

Educators can powerfully activate students' need to know content by launching a project with an "entry event" that engages interest and initiates questioning. An entry event can be almost anything: a video, a lively discussion, a guest speaker, a field trip, or a piece of mock correspondence that sets up a scenario.

A Driving Question

A good driving question captures the heart of the project in clear, compelling language, which gives students a sense of purpose and challenge. The question should be provocative, open-ended, complex, and linked to the core of what you want students to learn.

Student Voice and Choice

This element of project-based learning is key. In terms of making a project feel meaningful to students, the more voice and choice, the better. However, teachers should design projects with the extent of student choice that fits their own style and students.

21st Century Skills

A project should give students opportunities to build such 21st century skills as collaboration, communication, critical thinking, and the use of technology, which will serve them well in the workplace and life.

Inquiry and Innovation

Students find project work more meaningful if they conduct real inquiry, which does not mean finding information in books or websites.

Feedback and Revision

In addition to providing direct feedback, the teacher should coach students in using rubrics or other sets of criteria to critique one another's work.

A Publicly Presented Product

When students present their work to a real audience, they care more about its quality.

36.6 How Educators Can Support Project-Based Learning (PBL)

Educators can support project-based learning (PBL) as follows (Mergendoller and Thomas, 2001):

1. Time management—This theme relates to scheduling projects effectively by coordinating project schedules with other teachers.

2. Getting started—This theme is about orienting students, that is, getting them to think about the project well before they begin, giving them a rubric that clearly explains what they are expected to search for and try to accomplish, and jointly agreeing on grading criteria before the start of the project. The "getting started" theme is also about encouraging thoughtful work early on in the project in developing a research plan and a suitable research question while facilitating a sense of mission.

3. Establishing a culture that stresses student self-management—Here, responsibility is shifted from the teacher to students where they are involved in project design, they make decisions for themselves, and they are encouraged to learn how to learn.

4. Managing student groups—The emphasis is on establishing the appropriate grouping pattern, promoting full participation, and keeping track of each group's progress through discussion, monitoring, and recording evidence of progress.

5. Working with others outside the classroom, such as other teachers, parents, and people from the community, in order to work out the feasibility and nature of external partnerships.

6. Getting the most out of technological resources, such as judging the suitability of using technology for the project, making efficient use of the internet by being encouraged to make informed choices in exploring relevant web sites and developing critical thinking skills.

7. Assessing students and evaluating projects—This final theme refers, first, to the importance of grading students by using a variety of assessment methods, including individual and group grades, and giving emphasis to individual over group performance and, second, to adequately debriefing projects by demonstrating reflection strategies and collecting formative evaluation information from students about the project and how it might be improved.

36.7 Tips for Designing Effective Project-Based Learning (PBL)

Prepare the learning outcomes for the project-based learning (PBL).

Orientation is very important.

Select the questions, the project rationality, and importance.

Encourage students to search about the project.

Ask students to prepare a proposal about the project, and present it.

Provide feedback to students to improve their knowledge and skills.

Supervise the students, guiding them during the project.

Ask students to present their final project report, which includes findings, discussion, and so on.

Provide feedback to students to improve their knowledge and skills.

36.8 Tips for Implementing Online Project-Based Learning (PBL)

Project-based learning (PBL) can be implemented online with the help of new technologies.

The following steps are very important to implementing online project-based learning (PBL):

Check the technology facilities at your school, educators' homes, and students' homes.

Training about the technology used for online PBL is very important to pharmacy educators as well as students.

Prepare the learning outcomes for the project-based learning (PBL).

Orientation is very important.

Select the questions, the project rationality, and importance.

Encourage students to search about the project.

Ask students to prepare a proposal about the project, and present it.

Provide feedback to students to improve their knowledge and skills.

Supervise the students, guiding them during the project.

Ask students to present their final project report, which includes findings, discussion, and so on.

Provide feedback to students to improve their knowledge and skills.

36.9 Barriers to Implementing Online Project-Based Learning (PBL)

There are many barriers to implementing online project-based learning (PBL) as follows:

Lack of resources.

Lack of funds, financial issues.

Knowledge of pharmacy educators about PBL.

Attitude of pharmacy educators about PBL.

Attitude of pharmacy students about PBL.

Resistance of pharmacy educators towards PBL.

Resistance of pharmacy students towards PBL.

University/school culture towards PBL.

Lack of training about PBL.

Lack of motivation to train about PBL.

Lack of technologies to implement online PBL.

36.10 Conclusion

This chapter has discussed the history and importance of Project-Based Learning (PBL); the purpose of PBL; the process of PBL; tips for implementing effective PBL; tips for implementing effective online PBL; and barriers to conducting PBL. Project-based learning is designed to give students the opportunity to develop knowledge and skills through engaging projects set around challenges and problems they may face in the real world.

References

Keator, C.S., Vandre, D.D. and Morris, A.M., 2016. The challenges of developing a project-based self-directed learning component for undergraduate medical education. *Medical Science Educator*, 26(4), pp. 801–805.

Kokotsaki, D., Menzies, V. and Wiggins, A., 2016. Project-based learning: A review of the literature. *Improving Schools*, 19(3), pp. 267–277.

Mergendoller, J.R. and Thomas, J.W., 2001. *Managing project based learning: Principles from the field*. Buck Institute for Education.

Ms, A., 2010. *7essentials for project-based learning*. Educational Leadership.

37

Flipped Classes in Pharmacy Education

37.1 What Is a Flipped Classroom?

Persky and McLaughlin (2017) explained the definition of flipping the classroom as follows: flipping the classroom represents an ongoing paradigmatic shift in education from teacher-centered instructional strategies (e.g., lecturing) to learning-centered instructional strategies (e.g., active student engagement). The flipped classroom (also called reverse, inverse, or backwards classroom) is a pedagogical approach in which basic concepts are provided to students for pre-class learning so that class time can apply and build upon those basic concepts. While the term "flipped classroom" has garnered considerable attention in recent years, learner-centered pedagogies that effectively engage students in the learning process have a long and rich history. Approaches such as problem-based learning (PBL) and case-based learning (CBL) reflect many of the same learning-centered principles as flipped learning. However, the flipped classroom may vary from PBL or CBL in the level of guidance provided by an instructor and the types of activities used to facilitate learning, according to the following key flipped classroom elements: pre-class offloaded content, described as the process of packaging and delivering key foundational content to students prior to class; in-class active learning, represented by a wide range of well-developed and evidence-based, active-learning strategies; assessment, including diverse approaches to evaluating and providing feedback about student learning and holding students accountable for learning pre-class and in-class material; and exploration, characterized by student-initiated inquiry (Persky and McLaughlin, 2017; Tucker, 2012; Bergmann and Sams, 2012; Mok, 2014; Schmidt and Ralph, 2016).

37.2 History of the Flipped Classroom

The history of the flipped classroom goes back to 1984, when the first flipped classroom model was proposed. In the 1980s and 1990s, teachers in Russia

DOI: 10.1201/9781003230458-41

tried this instructional strategy. "[L]et pupils extract new things from autonomous reading of a textbook, which has been created accordingly. Allow them to consider it, then discuss it with their teacher at school and come to a united conclusion," Nechkina wrote of the flipped classroom. In 1993, Alison King published "From Sage on the Stage to Guide on the Side," in which she focuses on the importance of the use of class time for the construction of meaning rather than information transmission. While not directly illustrating the concept of "flipping" a classroom, King's work is often cited as an impetus for an inversion to allow for the educational space for active learning (Nechkina, 1984; Wikipedia, 2021; King, 1993). TBL is a model for flipping the classroom and has been used in medical education since the 1990s. The history of the flipped classroom in pharmacy education goes back to the 2000s.

37.3 Importance of the Flipped Classroom

The flipped classroom has many benefits for students such as the following:

Improves students' self-learning skills.

Improves students' understanding of the course materials.

Improves students' critical thinking skills.

Improves students' communication skills.

Improves students' attitude towards teamwork and collaboration.

Improves students' decision-making skills.

Improves students' time management skills.

Improves students' presentation skills.

Improves students' problem-solving skills.

Improves knowledge retention.

Improves leadership skills.

Improves ability to answer the cases.

Improves ability to provide/design pharmacist care and patient care.

Improves ability to identify and solve/prevent drug-related problems (DRPs).

Improves ability to provide good pharmacist care and patient care.

Provides more time for hands-on activities.

37.4 Purpose of the Flipped Classroom

The aim of the flipped classroom is to prepare students to be lifelong learners and improve their self-reading skills. Furthermore, to improve their basic and clinical knowledge and skills.

37.5 Reasons for the Flipped Classroom

Neilsen, 2012 reported the following reasons for flipping the classroom (Nielsen, 2012):

1. Many of our students don't have access to technology at home.
2. Flipped homework is still homework and there are a growing number of parents and educators who believe that mandatory homework needlessly robs children of their after-school time.
3. Flipping instruction might end up just providing more time to do the same type of memorization and regurgitation that just doesn't work.
4. If we really want transformation in education, one thing we must do is stop grouping students by date of manufacture, which the flipped classroom is ideally suited for. True flipping should include a careful redesign of the learning environment, but this is often overlooked.
5. The flipped classroom is built on a traditional model of teaching and learning: I lecture, you intake.

Millard (2012) reported the following reasons to flip the classroom (Millard, 2012):

1. Increases student engagement.
2. Strengthens team-based skills.
3. Offers personalized student guidance.
4. Focuses classroom discussion.
5. Provides faculty freedom.

37.6 Tips for Flipping the Classroom

Ash (2012) reported the following tips for flipping your classroom:

1. Don't get hung up on creating your own videos.
2. Be thoughtful about what parts of your class you decide to "flip" and when.
3. If possible, find a partner to create videos with.
4. Address the issue of access early.
5. Find a way to engage students in the videos.

37.7 Principles of the Flipped Classroom

Jeffries and Huggett (2014) reported the following principles:

- Students must have clear objectives for knowledge acquisition and access to
- materials that succinctly provide them with this information.
- The sources of assigned knowledge acquisition must be concise and focused to
- allow students to complete it before attending class.
- Students learn best in context; thus, assigned classwork should be focused on significant problems requiring application of their new knowledge and higher levels of learning.
- Peer teaching magnifies learning; thus, assigned classwork is best designed for and conducted in groups.
- Assessment drives learning; thus, assessment of class performance is desired and course assessments should mirror the higher level learning that has occurred in class, not merely the knowledge acquired in pre-class work.

37.8 Structure of the Flipped Classroom

Jeffries and Huggett, 2014 documented the following structure for the flipped classroom.

37.8.1 Pre-classroom Activities: Turning Your Lecture into Homework

Careful consideration is required when replacing lectures with flipped sessions. First and foremost, students must have clear objectives for the session as they approach their preparatory assignments. These objectives should be constructed using verbs associated with the new Bloom's taxonomy that describe what the student will be able to do at the end of the session. This way, students can prepare in a focused manner and not waste time guessing what will be covered.

37.8.2 Flipped Classroom Preparation

37.8.2.1 Setting Up the Room

The physical setting for the flipped classroom is likely the same room where you have conducted your lectures. Ideally this would be in a room with a flat floor, with tables and movable chairs to allow students to configure themselves into learning groups. Often, the flipped session must take place in a tiered classroom that is better suited for viewing a lecture than participating in a group activity.

37.8.2.2 Design of the Session

One of the best features of designing a flipped learning experience is that there are many types of activities that can be employed to hold student interest and contribute to learning. Prior to beginning the activity, you should ask several relevant questions:

- Assuming that students have met the knowledge objectives in the preparatory session, what will you expect the students to do with that information? This forms the basis for writing the session objectives.

- How will you assure readiness and participation? Students who do not complete the assigned work will not be effective participants in classroom work, so an effective session must deal with this question.

- What activities will be used and how will they be applied toward meeting the objectives? In the next section, a variety of possible activities are introduced that can be used to stimulate individual or group learning.

- How will learning be assessed? Assessment of learning plays an important part in the design of the session. Holding students accountable for advanced stages of their learning is essential to making the flipped classroom successful.

37.8.3 Readiness Assurance

Readiness assurance is a term borrowed from team-based learning (Al-Meman et al., 2014):

- IRAT (individual readiness assurance test): in this step, the students will take the exam individually and solve it in 15 minutes as an example; you may adjust it based on the course and need. The type of questions could be multiple-choice questions (MCQs) as this is reported or, in my opinion, the lecturer may change it to any type such as short essay or any form. The most important issue is to achieve the course learning outcome and to improve students' knowledge and skills. However, 5 to 15 MCQs could be an option for 10 to 15 minutes.

- gRAT/tRAT (group or team readiness assurance test): in this step the same iRAT questions are redistributed to all groups for discussion, answering, and justifying the answers. The time allowed for gRAT is 30 minutes. The students are encouraged to vote for the best correct answer during the gRAT if there is no agreement, to enhance discussion and interaction among the group members.

- The next 20 minutes are considered for the in-place presentation of the answers. The presentation of the answer must be done simultaneously among the groups (at the same time) by voting to make it more interactive, followed by justifying and defending against the answers.

- The next 20 minutes are allocated for open discussion.

- The last 10 minutes are assigned for peer evaluation and appeals. Encourage students to evaluate each teammate on their contributions to the team's success and their own learning. Developing an evaluation checklist with both qualitative and quantitative questions could be easier and facilitate the TBL session.

37.8.4 Homework in Class and Related Techniques

37.8.4.1 Worksheet

Suggestions for application:

1. Do not make this the entire session, as students get bored fairly quickly.

2. Distribute the worksheet during class—if you hand it out in advance (or if students have access to it from previous students), students will avoid the session.

3. Allow students to complete it as a group. This allows students with a better grasp of the material to teach those who need help.
4. Combine the worksheet with adapted classroom assessment techniques (see later) to maximize the benefit.
5. Allow each group to submit one worksheet for credit, or give a short quiz at the end of the session containing similar problems to ensure that all students have achieved session objectives.

37.8.4.2 Dry Lab

Suggestions for application:

1. Distribute the dry lab data during class.
2. Have students work in small groups to complete assigned labs.
3. Have students submit completed group work before presentations to ensure that everyone participates.
4. Vary lab material from year to year to ensure students won't use material and answers from previously enrolled students.
5. Assess achievement of objectives with a short set of relevant problems at the end of, or shortly after, the session.

37.8.4.3 Case report

37.8.4.4 Review session

37.8.4.5 Others

Assessment of Student Progress

The grading system for the TBL session allocates marks for each step. Example: 10 marks for iRAT, 15 marks for gRAT, 5 marks for the peer evaluation, 20 marks for the cases, or you may adjust it as needed. Plus other assessment methods.

37.8.5 Tips for Designing the Effective Flipped Classroom

Prepare the learning outcomes for the flipped classroom.

Orientation is very important.

Select the questions, the project rationale, and importance.

Follow the structure of the flipped classroom as previously explained.

Provide feedback to students to improve their knowledge and skills.

37.9 Tips for Implementing the Online Flipped Classroom

The flipped classroom can be implemented online with the help of new technologies.

The following steps are very important to implement the flipped classroom online:

Check the technology facilities at your school, educators' homes, and students' homes.

Training about the technology use for the online flipped classroom is very important to pharmacy educators as well as students.

Orientation is very important.

Select the questions, the project rationale, and importance.

Follow the structure of flipped classroom as previously explained.

Provide feedback to students to improve their knowledge and skills.

Prepare for online readiness (iRAT and gRAT):

- Online iRAT (individual readiness assurance test) by using Blackboard or Moodle as well as Microsoft Teams or any video conferencing platform to monitor the iRAT. For this step you may make one Microsoft Teams link for all students. In this step, the students will take the exam individually and solve it in 15 minutes as an example; you may adjust it based on the course and need. The type of questions could be multiple-choice questions (MCQs) as this is reported or, in my opinion, the lecturer may change it to any type such as short essay or any form. The most important issue is to achieve the course learning outcomes and to improve students' knowledge and skills. However, 5 to 15 MCQs could be an option for 10 to 15 minutes.

- Online gRAT/tRAT (group or team readiness assurance test) by using Blackboard or Moodle as well as Microsoft Teams or any video conferencing platform to monitor the gRAT. At this step you may make one Microsoft Teams link for all students and distribute the teams through Microsoft Teams. In this step the same iRAT questions are redistributed to all groups for discussion, answering, and justifying the answers. The time allowed for gRAT is 30 minutes. Students are encouraged to vote for the best correct answer during the gRAT if there is no agreement, to enhance discussion and interaction among the group members.

- The next 20 minutes are considered for the in-place presentation of the answers. At this step you may make one Microsoft Teams link for all students. The presentation of the answer must be done simultaneously among the groups (at the same time) by voting to make it more interactive, followed by justifying and defending against the answers.
- The next 20 minutes are allocated for open discussion.

37.10 Barriers to Implementing the Flipped Classroom

There are many barriers to implementing the flipped classroom as follows:

Lack of resources.

Lack of funds, financial issues.

Knowledge of pharmacy educators about the flipped classroom.

Attitude of pharmacy educators towards the flipped classroom.

Attitude of pharmacy students towards the flipped classroom.

Resistance of pharmacy educators towards the flipped classroom.

Resistance of pharmacy students towards the flipped classroom.

University/school culture towards the flipped classroom.

Lack of training about the flipped classroom.

Lack of motivation to train about the flipped classroom.

Lack of technologies to implement the online flipped classroom.

37.11 Conclusion

This chapter has discussed the history and importance of the flipped classroom; the purpose of the flipped classroom; the process of the flipped classroom; tips for implementing an effective flipped classroom; tips for implementing an effective online flipped classroom; and barriers to conducting a flipped classroom.

References

Al-Meman, A., Al-Worafi, Y.M. and Saeed, M.S., 2014. Team-based learning as a new learning strategy in pharmacy college, Saudi Arabia: Students' perceptions. *Universal Journal of Pharmacy*, 3(3), pp. 57–65.

Ash, K., 2012. Educators view flipped model with a more critical eye. *Education Week*, 32(2), pp. S6–S7.

Bergmann, J. and Sams, A., 2012. *Flip your classroom: Reach every student in every class every day*. International society for technology in education.

Jeffries, W.B. and Huggett, K.N., 2014. Flipping the classroom. In *An introduction to medical teaching* (pp. 41–55). Springer.

King, A., 1993. From sage on the stage to guide on the side. *College Teaching*, 41(1), pp. 30–35.

Millard, E., 2012. 5 reasons flipped classrooms work. *University Business Magazine*. Available at: https://margopolo04.files.wordpress.com/2013/12/flipped-classroom-reasons.pdf

Mok, H.N., 2014. Teaching tip: The flipped classroom. *Journal of Information Systems Education*, 25(1), p. 7.

Nielsen, L., 2012. Five reasons I'm not flipping over the flipped classroom. *Technology & Learning*, 32(10), pp. 46–46.

Nechkina, Militsa., 1984. Increasing the effectiveness of a lesson. *Communist*, 2, p. 51.

Persky, A.M. and McLaughlin, J.E., 2017. The flipped classroom—from theory to practice in health professional education. *American Journal of Pharmaceutical Education*, 81(6).

Schmidt, S.M. and Ralph, D.L., 2016. The flipped classroom: A twist on teaching. *Contemporary Issues in Education Research* (CIER), 9(1), pp. 1–6.

Tucker, B., 2012. The flipped classroom. *Education Next*, 12(1), pp. 82–83.

Wikipedia., 2021. *Flipped classroom*. Available at: https://en.wikipedia.org/wiki/Flipped_classroom

38

Educational Games in Pharmacy Education

38.1 What Are Educational Games?

Cook (2014) defined educational games as: educational games are activities with rules and a defined outcome (winning, losing) or other feedback (e.g., points) that facilitate comparisons of performance. Games typically have explicit goals and a compelling storyline, and thus have the potential to engage learners and encourage their continued practice with the objective of improved knowledge and skill acquisition and application. However, the benefits of online educational games in medical education are still largely hypothetical, with only a few descriptions and even fewer comparative studies (Cook, 2014). The gamification of learning is an educational approach that seeks to motivate students by using video game design and game elements in learning environments. The goal is to maximize enjoyment and engagement by capturing the interest of learners and inspiring them to continue learning. Gamification, broadly defined, is the process of defining the elements which comprise games, make those games fun, and motivate players to continue playing, then using those same elements in a non-game context to influence behavior. There are two forms of gamification: structural, which means no changes to subject matter, and the altered content method that adds subject matter (Kapp, 2012; Hsin-Yuang et al., 2013; Wikipedia, 2021).

38.2 History of Educational Games

The history of educational games goes back to the 1960s when the first real educational game was Logo Programming. Turtle Academy released Logo Programming in 1967 with the intent of teaching people how to program using the LOGO programming language. It also happened to serve as a tool to learn mathematical concepts. The next educational game didn't arrive until 1973. Lemonade Stand was created as a business simulation game and taught players basic economics; however, it was reported that games are as old as human beings; the history of educational games was first detected at the time when Socrates and Plato used a kind of verbal play in their "dialogues." In the 19th century, Froebel integrated "learning," "game," and "play," and, in the 1990s, digital games were developed further and became dramatically widespread among youth, resulting in a cohort of incoming students accustomed to digital game plays (Bigdeli and Kaufman, 2017). The history of educational games in pharmacy education goes back to the 1980s (Oliver et al., 1995; McVey et al., 1989).

Pharmacy educators in other fields have been using simulation games to sensitize their students to concerns facing future clients. One of these, variously known as "Into Aging" and the "Aging Game," has documented success in introducing nursing and medical students to the problems confronted by the frail elderly and is making its way into the required curricula at medical schools. While "Into Aging" in one of its present formats could be used in pharmacy education, only two of the more than 60 life events in the game have to do with medication. Moreover, the focus of "Into Aging" is on the frail elderly and institutional care. Given that close to 95% of Americans over 65 live outsides of institutions, their inability to manage medications at home has been reported as a primary cause for nearly one-quarter of nursing-home admissions (Oliver et al., 1995; McVey et al., 1998).

38.3 Importance of Educational Games

Game-based learning (GBL) has many benefits as follows (Aburahma and Mohamed, 2015):

> By playing games, students become more motivated to learn, pay attention, and participate in set tasks.
>
> Games help students to become a part of a team as well as take responsibility for their own learning. They can also be a great classroom management tool, helping to motivate a class.
>
> Improves competition among students.
>
> Relieves stress among students.
>
> Fun.
>
> New knowledge.
>
> Has cognitive benefits.
>
> Has motivational benefits.
>
> Has emotional benefits.
>
> Has social benefits.

38.4 Purpose of Educational Games

The aim of educational games is to prepare students to convey knowledge. When employed properly, educational games build knowledge and skills and are enjoyable for the participants and appeal to students' competitive nature, which motivates them to play the game. Educational games often promote higher-level discussions which help to enhance the communication, social interaction, and critical-thinking skills essential in health care. The games also allow health care educators to create real-life scenarios without real-life consequences. The format of educational games creates a setting that decreases student stress and facilitates student learning (Barclay et al., 2011).

38.5 Educational Games in Pharmacy Education

Aburahma and Mohamed (2015) conducted a literature review about educational games in pharmacy education and reported the following examples (Aburahma and Mohamed, 2015):

Who Wants to Be a Med Chem Millionaire?

Six teams (each of six members) sit in front of the classroom while nonplaying students remain in the classroom as "studio audience." Game questions are projected on screens and teams ring a bell to answer a question. Play time is 45 minutes. The team with the most Med Chem Moolah (play money) at the end of their session wins.

Prizes/rewards: Money donated to charity named by the winning team (Roche et al., 2004).

Who Wants to Be a Millionaire? Five

Multiple-choice questions are presented to students using PowerPoint. The first student to raise his/her hand is selected by the instructor to answer the question. Students can be assisted by a friend in the room, or by audience help via a poll. Game time is approximately 5 minutes and is followed by a lecture.

Prizes/rewards: candy (Grady et al., 2013).

Jeopardy

The class is randomly assigned to 16 teams, each of approximately 10 students. The teams sit in assigned areas in the classroom. A student from the audience selects a question to display on a PowerPoint projection overhead. Any group can participate by raising a hand to answer the question. Correct answers are awarded the appropriate points, and that team selects the next question. If a group answers incorrectly, a second team attempts to answer and earn the points. Game time is approximately 30 minutes, followed by a 45-minute lecture.

Prizes/rewards: extra credit point to the group with the highest score (Grady et al., 2013).

Survivor

Teams of 20 students are asked three major questions. They take 5 minutes to discuss the question within their groups. One member of each team writes the answers on a whiteboard at the front of the lecture hall. The team with the largest number of correct answers proceeds to the next question. This cycle is repeated for all three questions. Game time is 1.5 hours (Grady et al., 2013).

Crossword Puzzle

The whole class participates in solving a 5-minute crossword puzzle containing information presented in the lecture (Shah et al., 2010).

Race to Glucose

Students in groups of five or six roll a die and move game pieces along the gluconeogenesis pathway while addressing questions and changes in physiological conditions. The team who finishes the pathway first wins. Game time is approximately 2 hours on two consecutive days (Rose, 2011).

Medication Mysteries Infinite Case Tool (MMICT)

During a 2-hour laboratory session, groups of three students are provided with the MMICT packet containing a game board, decks of drug, confusion, and personality cards, a 6-sided die, instruction sheet, patient demographic sheets, and an evaluation rubric. Each student assumes a role: patient, pharmacist, or evaluator (Sando et al., 2013).

Bingo Game

The game is composed of a 5 x 5 grid with total of 25 squares, each containing an activity to encourage students to review course material (online self-quizzes), to motivate students to perform better on graded activities (examinations and competencies), to appeal to students with different learning styles (posters, computer animations, videos, crossword puzzles), and to encourage close attention to required material (identify errors in textbook or class). Students who achieve bingo (5 squares in a row vertically, horizontally, or corner-to-corner) earn a 5-point (5%) bonus added to the final course grade.

Prizes/rewards: 5% added to final course grade (Tietze, 2007).

PK Poker

The class is divided into 13 groups of approximately 10 students. Each group starts with a $500 bankroll and places a bet on their ability to answer a question correctly. Students have 2 minutes to respond to each question. Game time is 50 minutes for two class periods.

Prizes/rewards: bonus points added to the total points of the course (Persky et al., 2007).

Pharmacy Scene Investigation (PSI)

The class is divided into six groups of approximately 22 students with two members acting as lead detectives. The game presents an unsolved death scenario about an individual found dead with initial indications of suicide and multiple potential murderer suspects. The game is presented divided into one 50-minute class to play and one 50-minute class to debrief (Persky et al., 2007).

Clue Game (CG)

Game is based on a murder mystery. Each student in a five-member group researches four different drugs from the Top 300 drugs, then teaches them to other group members. Students receive clues to determine the murderer (e.g., physician who prescribed a medication with severe adverse effects), weapon used, and location. If the student answers correctly, the group successfully completes the game. If the student does not, the team is disqualified. Game time is one 50-minute class period to play and one 50-minute class to debrief (Persky et al., 2007).

38.6 Tips for Designing Effective Game-Based Learning (GBL)

Prepare the learning outcomes for the game-based learning (GBL); this could be different from one course to another.

Select the type of game, adapt it or design it.

Training is very important for educators to succeed in game-based learning.

Orientation is very important.

Explain the theory part to students if needed or ask them to read it before the game.

Provide feedback to students to improve their knowledge and skills.

Allocate rewards, prizes to motivate the students.

38.7 Tips for Implementing Online Game-Based Learning (GBL)

Game-based learning (GBL) can be implemented online.

The following steps are very important to implementing game-based learning (GBL) online:

Check the technology facilities at your school, educators' homes and students' homes.

Training about the technology used for online game-based learning (GBL) is very important to pharmacy educators as well as students.

Prepare the learning outcomes for the GBL; this could be different from one course to another.

Select the type of game, adapt it or design it.

Training is very important for educators to succeed in online GBL.

Orientation is very important.

Explain the theory part to students if needed or ask them to read it before the game.

Provide feedback to students to improve their knowledge and skills.

Allocate rewards, prizes to motivate the students.

38.8 Barriers to Implementing Game-Based Learning (GBL)

There are many barriers to implementing game-based learning (GBL) as follows:

Lack of resources.

Lack of funds, financial issues.

Knowledge of pharmacy educators about GBL.

Attitude of pharmacy educators towards GBL.

Attitude of pharmacy students towards GBL.

Resistance of pharmacy educators towards GBL.

Resistance of pharmacy students towards GBL.

University/school culture towards GBL.

Lack of training about GBL.

Lack of motivation to train about GBL.

Lack of technologies to implement online GBL.

38.9 Conclusion

This chapter has discussed the history and importance of educational games; the purpose of games in education; examples of games in pharmacy education; and the barriers to implementing game-based learning (GBL) in pharmacy education.

References

Aburahma, M.H. and Mohamed, H.M., 2015. Educational games as a teaching tool in pharmacy curriculum. *American Journal of Pharmaceutical Education*, 79(4).

Barclay, S.M., Jeffres, M.N. and Bhakta, R., 2011. Educational card games to teach pharmacotherapeutics in an advanced pharmacy practice experience. *American Journal of Pharmaceutical Education*, 75(2).

Bigdeli, S. and Kaufman, D., 2017. Digital games in medical education: Key terms, concepts, and definitions. *Medical Journal of the Islamic Republic of Iran*, 31, p. 52.

Cook, D.A., 2014. Teaching with technological tools. In *An introduction to medical teaching* (pp. 123–146). Springer.

Grady, S.E., Vest, K.M. and Todd, T.J., 2013. Student attitudes toward the use of games to promote learning in the large classroom setting. *Currents in Pharmacy Teaching and Learning*, 5(4), pp. 263–268.

Hsin-Yuan Huang, W. and Soman, D., 2013. *A practitioner's guide to gamification of education*. Rotman School of Management, University of Toronto.

Kapp, K.M., 2012. *The gamification of learning and instruction: Game-based methods and strategies for training and education*. John Wiley & Sons.

McVey, L.J., Davis, D.E. and Cohen, H.J., 1989. The 'aging game': An approach to education in geriatrics. *JAMA*, 262(11), pp. 1507–1509.

Oliver, C.H., Hurd, P.D., Beavers, M., Gibbs, E., Goeckner, B. and Miller, K., 1995. Experiential learning about the elderly: The geriatric medication game. *American Journal of Pharmaceutical Education*, 59(2), pp. 155–157.

Persky, A.M., Stegall-Zanation, J. and Dupuis, R.E., 2007. Students perceptions of the incorporation of games into classroom instruction for basic and clinical pharmacokinetics. *American Journal of Pharmaceutical Education*, 71(2).

Roche, V.F., Alsharif, N.Z. and Ogunbadeniyi, A.M., 2004. Reinforcing the relevance of chemistry to the practice of pharmacy through the who wants to be a med chem millionaire? Learning game. *American Journal of Pharmaceutical Education*, 68(5), p. 116.

Rose, T.M., 2011. A board game to assist pharmacy students in learning metabolic pathways. *American Journal of Pharmaceutical Education*, 75(9).

Sando, K.R., Elliott, J., Stanton, M.L. and Doty, R., 2013. An educational tool for teaching medication history taking to pharmacy students. *American Journal of Pharmaceutical Education*, 77(5).

Shah, S., Lynch, L.M. and Macias-Moriarity, L.Z., 2010. Crossword puzzles as a tool to enhance learning about anti-ulcer agents. *American Journal of Pharmaceutical Education*, 74(7).

Tietze, K.J., 2007. A bingo game motivates students to interact with course material. *American Journal of Pharmaceutical Education*, 71(4).

Wikipedia., 2021. *Gamification of learning*.

39

Web-Based Learning in Pharmacy Education

39.1 Terminologies

Web-based learning is often called online learning or e-learning because it includes online course content. Discussion forums via email, videoconferencing, and live lectures (videostreaming) are all possible through the Web. Web-based courses may also provide static pages such as printed course materials. One of the values of using the Web to access course materials is that Web pages may contain hyperlinks to other parts of the Web, thus enabling access to a vast amount of Web-based information. A "virtual" learning environment (VLE) or managed learning environment (MLE) is an all-in-one teaching and learning software package. A VLE typically combines functions such as discussion boards, chat rooms, online assessment, tracking of students' use of the Web, and course administration. VLEs act as any other learning environment in that they distribute information to learners. VLEs can, for example, enable learners to collaborate on projects and share information. However, the focus of Web-based courses must always be on the learner—technology is not the issue, nor necessarily the answer (McKimm et al., 2003). Web-based learning refers to the type of learning that uses the internet as an instructional delivery tool to carry out various learning activities. It can take the form of a pure online learning in which the curriculum and learning are implemented online without face-to-face meeting between the instructor and the students, or a hybrid in which the instructor meets the students half of the time online and half of the time in the classroom, depending on the needs and requirements of the curriculum. Web-based learning can be integrated into a curriculum that turns into a full-blown course or as a supplement to traditional courses. Non face to face, using Web technologies, it is learning that occurs with lessons conducted via the internet via internet-enabled transfer of knowledge and skills. It is a type of instructional strategy that is based on resources available on the Web. Learners may learn individually or in groups. This involves discussions, forums, and course content delivered through emails, live lectures, and/or videos. Teaching material is presented via the internet, specifically

DOI: 10.1201/9781003230458-43

the World Wide Web. Web-based learning consists of instruction programs using attributes and resources of the Web to create a meaningful learning and interactive environment; it is used synonymously as online learning. It is a form of computer-based instruction that uses the World Wide Web as the primary delivery method of information; a type of learning employing Web-based technologies. It may concern individuals or learning communities. It involves the manipulation of a set of content analysis techniques aiming to establish a conceptual model of a task-specific domain. Learning via the internet is a form of education where the resources and student-faculty and student-student interactions take place on the world wide web (Kidd and Song); Tsai and Machado (2002).

E-learning is mostly associated with activities involving computers and interactive networks simultaneously. The computer does not need to be the central element of the activity or provide learning content. However, the computer and the network must hold a significant involvement in the learning activity (Tsai and Machado, 2002).

Web-based learning is associated with learning materials delivered in a Web browser, including when the materials are packaged on CD-ROM or other media (Tsai and Machado, 2002).

Online learning is associated with content readily accessible on a computer. The content may be on the Web or the internet, or simply installed on a CD-ROM or the computer hard disk (Tsai and Machado, 2002).

Distance learning involves interaction at a distance between instructor and learners and enables timely instructor reaction to learners. Simply posting or broadcasting learning materials to learners is not distance learning. Instructors must be involved in receiving feedback from learners. For each of these concepts, the discriminating feature must be the primary characteristic of the learning activity. Intensive use of the feature is required, since incidental or occasional use of a characteristic feature is not sufficient to qualify for a certain type of learning. For instance, running a CBT application from a file-server does not qualify as E-learning; and e-mailing a teacher after taking a class on campus is not sufficient to qualify as distance learning (Tsai and Machado, 2002).

The term *Web 2.0* refers to an evolving number of Web applications that are both open and social in nature (Cain and Fox, 2009).

39.2 History of Web-Based Learning

The origin of the internet began during the Cold War circa 1969 in the United States of America. A network was created to link all the military computers across the US so that in the event of a nuclear war, the American military equipment could still function. Within 15 years of the creation of this network, the connected sites were only military and academic sites. In 1972, there were 40 small networks connected to the ARPAnet. At that time, this network was used to send small text files among users. Now this is known as e-mail and is widely used. In the 1980s, these networks were connected commercially. We were able to visit from one network to a different network and by this time the term "internet" was used (Abd Rashid et al., 2016) The internet is the catalyst to WBL. There are three initial generations of web support in practice as follows (Ahamer, 2010):

First generation 1999: Three cover pages representing content delivered to students via both a web platform and a paper manuscript: "Technology Assessment" (TA), "Systems Analysis and Biology" (SB), and "Environmental Technology" (ET) at FH Joanneum over a period of six years.

Second generation 2002: Time structure of eight face-to-face meetings with online phases in between. Only one real meeting was replaced by a virtual one (cloud).

Third generation 2003–2005: Welcome screen of SurfingGlobalChange (SGC).

39.3 Advantages of Web-Based Learning

Web-based learning (WBL) has many advantages as follows (McKimm et al., 2003):

- Ability to link resources in many different formats.
- Can be an efficient way of delivering course materials.
- Resources can be made available from any location and at any time.
- Potential for widening access—for example, to part-time, mature, or work-based students.
- Can encourage more independent and active learning.
- Can provide a useful source of supplementary materials to conventional programs.
- Cost-effectiveness.
- Flexibility.
- Students can receive quick feedback on their performance.
- Useful for self-assessments—for example, multiple-choice questions.
- A convenient way for students to submit assessments from remote sites.
- Computer marking is an efficient use of staff time.

Web-based learning has several potential benefits within clinical education (Isaacs et al., 2019):

- The augmentation of traditional clinical teaching, decreasing educator time commitments for direct instruction with students.
- The provision of standardized learning experiences across geographically diverse settings.
- The remediation of student knowledge.
- The promotion of self-directed learning.

39.4 Disadvantages of Web-Based Learning

Web-based learning (WBL) has the following disadvantages (McKimm et al., 2003):

- Access to appropriate computer equipment can be a problem for students.
- Learners find it frustrating if they cannot access graphics, images, and video clips because of poor equipment.
- The necessary infrastructure must be available and affordable.
- Information can vary in quality and accuracy, so guidance and signposting is needed.
- Students can feel isolated.
- Most online assessment is limited to objective questions.
- Security can be an issue.
- It can be difficult to authenticate students' work.
- Computer marked assessments tend to be knowledge based and measure surface learning.

39.5 Purpose of Web-Based Learning

The aim of Web-based learning (WBL) is to provide and facilitate the teaching and learning cycle, which allows students to learn from anywhere.

39.6 Tips for Designing Effective Web-Based Learning (WBL)

Cook and Dupras (2004) reported the following steps for preparing effective Web-based learning:

1. Perform a needs analysis and specify goals and objectives.
2. Determine your technical resources and needs.
3. Evaluate preexisting software and use it if it fully meets your needs.
4. Secure commitment from all participants and identify and address potential barriers to implementation.
5. Develop content in close coordination with website design:
   - Capitalize on the unique capabilities of the Web by appropriately using multimedia, hyperlinks, and online communication.
   - Adhere to principles of good webpage design.
   - Prepare a timeline, plan for up-front time investment.
6. Encourage active learning—self-assessment, reflection, self-directed learning, problem-based learning, learner interaction, and feedback.
7. Facilitate and plan to encourage use by the learner:
   - Make the website accessible and user friendly.
   - Provide time for learning.
   - Motivate and remind; consider rewards and/or consequences.
8. Evaluate—both learners and the course.
9. Pilot the website before full implementation.
10. Plan to monitor online communication and maintain the site by resolving technical problems, periodically verifying hyperlinks, and regularly updating content.

39.7 Barriers to Web-Based Learning (WBL)

There are many barriers to implementing Web-based learning (WBL) as follows:

Lack of resources.
Lack of funds, financial issues.

Knowledge of pharmacy educators about WBL.

Attitude of pharmacy educators towards WBL.

Attitude of pharmacy students towards WBL.

Resistance of pharmacy educators towards WBL.

Resistance of pharmacy students towards WBL.

University/school culture towards WBL.

Lack of training about WBL.

Lack of motivation to train about WBL.

Lack of technologies to implement WBL.

39.8 Conclusion

This chapter has discussed the history and importance of Web-based learning (WBL); the purpose of WBL; advantages and disadvantages of WBL; tips for implementing effective WBL; and barriers to conducting WBL.

References

Abd Rashid, Z., Kadiman, S., Zulkifli, Z., Selamat, J., Hisyam, M. and Hashim, M., 2016. Review of web-based learning in TVET: history, advantages and disadvantages. *International Journal of Vocational Education and Training Research*, 2(2), p. 7.

Ahamer, G., 2010. A short history of web based learning including GIS. *International Journal of Computer Science & Emerging Technologies*, 1(4), pp. 101–111.

Cain, J. and Fox, B.I., 2009. Web 2.0 and pharmacy education. *American Journal of Pharmaceutical Education*, 73(7).

Cook, D.A. and Dupras, D.M., 2004. A practical guide to developing effective web-based learning. *Journal of General Internal Medicine*, 19(6), pp. 698–707.

Isaacs, A.N., Walton, A.M., Gonzalvo, J.D., Howard, M.L. and Nisly, S.A., 2019. Pharmacy educator evaluation of web-based learning. *The Clinical Teacher*, 16(6), pp. 630–635.

McKimm, J., Jollie, C. and Cantillon, P., 2003. ABC of learning and teaching: Web based. *BMJ*, 326, pp. 870–873.

Tsai and Machado, 2002. E-learning basics: Essay: E-learning, online learning, web-based learning, or distance learning: Unveiling the ambiguity in current terminology. *eLearn*, 7, p. 3.

40

Lecture-Based/Interactive Lecture-Based Learning in Pharmacy Education

40.1 What Is Lecture-Based Learning (LBL)?

Lecture-based learning (LBL) can be defined as that type of learning where the educator/teacher educates/teaches students by delivering presentations and PowerPoint presentations (lectures) verbally to the students and learners.

40.2 What Is Interactive Lecture-Based Learning (ILBL)?

Interactive lecture-based learning (ILBL) can be defined as that type of learning where the educator/teacher educates/teaches students by delivering presentations and PowerPoint presentations (lectures) verbally to the students and learners in addition to using other active teaching strategies such as videos and cases delivered to the students and learners which engages the students and motivates them to participate actively during the class.

40.3 History of Lecture-Based Learning (LBL) and Interactive Lecture-Based Learning (ILBL)

The history of lecture-based learning (LBL) goes back to the history of teaching and learning and developed throughout history. Palincsar, a professor at the University of Michigan, put forward the concept of interactive teaching in the classroom in the 1970s, which is now called interactive learning or interactive teaching (Huang and Liu, 2014).

DOI: 10.1201/9781003230458-44

40.4 Advantages of Lecture-Based Learning

Literature summarizes the advantages of lecture-based learning (LBL) as follows (DoingCL, 1997; Cashin, 1985; Bonwel, 1996):

- Effective lecturers can communicate the intrinsic interest of a subject through their enthusiasm.
- Lectures can present material not otherwise available to students.
- Lectures can be specifically organized to meet the needs of particular audiences.
- Lectures can present large amounts of information.
- Lectures can be presented to large audiences.
- Lecturers can model how professionals work through disciplinary questions or problems.
- Lectures allow the instructor maximum control over the learning experience.
- Lectures present little risk for students.
- Lectures appeal to those who learn by listening.

40.5 Disadvantages of Lecture-Based Learning

Literature summarizes the disadvantages of lecture-based learning (LBL) as follows (DoingCL, 1997; Cashin, 1985; Bonwel, 1996):

- Lectures fail to provide instructors with feedback about the extent of student learning.
- In lectures students are often passive because there is no mechanism to ensure that they are intellectually engaged with the material.
- Students' attention wanes quickly after 15 to 25 minutes.
- Information tends to be forgotten quickly when students are passive.
- Lectures presume that all students learn at the same pace and are at the same level of understanding.
- Lectures are not suited for teaching higher orders of thinking such as application, analysis, synthesis, or evaluation; for teaching motor skills; or for influencing attitudes or values.
- Lectures are not well suited for teaching complex, abstract material.

- Lectures requires effective speakers.
- Lectures emphasize learning by listening, which is a disadvantage for students who have other learning styles.

40.6 Purpose of Interactive Lecture-Based Learning

The aim of interactive lecture-based learning (ILBL) is to engage the students and motivate them to participate actively during the class.

40.7 Importance of Interactive Lecture-Based Learning

Improve students' engagement in the class.
Improve students understanding of the course materials.
Improve students' critical thinking skills.
Improve students' communication skills.
Improve students' attitude towards teamwork and collaboration.
Improve students' decision-making skills.
Improve students' time management skills.
Improve students' presentation skills.
Improve students' problem-solving skills.
Improve knowledge retention.
Improve leadership skills.
Improve ability to answer the cases.
Improve ability to provide/design pharmacist care and patient care.
Improve ability to identify and solve/prevent drug-related problems (DRPs).

40.8 Tips for Designing Effective Interactive Lecture-Based Learning

Orientation is very important for the success of interactive lecture-based learning. Orientation should be about the course objectives, learning outcomes, topics, educational resources, and assessment.

Provide the students with the course resources such as books, lecture notes, and others.

Prepare the facilities for interactive lecture-based learning.

Prepare the lecture presentations and PowerPoint presentations (PPT) in the best effective way: simple and clear, with clear color, clear content, limited lines in each slide; include lecture learning outcomes and outline, link it with videos and other interactive tools, and add references.

Allocate enough time for the interaction between you and students.

Use more than one interactive tool such as videos, cases, and so on.

Engage students in the lecture.

40.9 Tips for Implementing Online Interactive Lecture-Based Learning

Online interactive lecture-based learning and online lectures or online interactive lectures can be implemented online as follows:

Check the technology facilities at your school as well as at students' homes.

Training about technology use for the online lectures is very important to pharmacy educators as well as students.

Orientation is very important for the success of online interactive lecture-based learning.

Provide students with the online course resources such as e-books, lecture notes, and others.

Make mock online interactive lecture-based sessions and get feedback from all students and peers.

40.10 Barriers to Interactive Lecture-Based Learning

There are many barriers to implementing interactive lecture-based learning as follows:

Lack of resources.

Lack of funds, financial issues.

Knowledge of pharmacy educators about interactive lecture-based learning.

Attitude of pharmacy educators about interactive lecture-based learning.

Attitude of pharmacy students about interactive lecture-based learning.

Resistance of pharmacy educators towards interactive lecture-based learning.

Resistance of pharmacy students towards interactive lecture-based learning.

University/school culture towards interactive lecture-based learning.

Lack of training about interactive lecture-based learning.

Lack of motivation to train about interactive lecture-based learning.

Lack of technologies to implement interactive lecture-based learning.

40.11 Conclusion

This chapter has discussed the history and importance of interactive lecture-based learning; advantages and disadvantages of lectures; the purpose and importance of interactive lecture-based learning (ILBL); steps for effective ILBL; and barriers to ILBL.

References

Bonwell, C.C., 1996. Enhancing the lecture: Revitalizing a traditional format. In Sutherland, T.E. and Bonwell, C.C., eds. *Using active learning in college classes: A range of options for faculty.* New Directions for Teaching and Learning No. 67.

Cashin, W.E., 1985. *Improving lectures.* Idea Paper No. 14. Kansas State University, Center for Faculty Evaluation and Development.

Doing, CL., 1997. *Advantages and disadvantages of lectures.* Available at: wceruw.org

Huang, M.Y. and Liu, Y.Z., 2014, November. *Operation models of interactive learning.* In *International Conference on Social Science* (ICSS 2014) (pp. 256–260). Available at: https://www.atlantis-press.com/proceedings/icss-14/14620

41

Blended Learning in Pharmacy Education

41.1 What Is Blended Learning (BL)?

Blended learning can be defined as follows (Graham, 2006; Garrison and Kanuka, 2004; Garrison and Vaughan, 2008; Hrastinski, 2019; Bersin, 2004): Blended learning is an approach to education that combines online educational materials and opportunities for interaction online with traditional place-based classroom methods. It requires the physical presence of both teacher and student, with some elements of student control over time, place, or path. Blended learning systems combine face-to-face instruction with computer-mediated instruction. Garrison and Kanuka (2004) define blended learning as "the thoughtful integration of classroom face-to-face learning experiences with online learning experiences." Thus, we can conclude that there is general agreement that the key ingredients of blended learning are face-to-face and online instruction or learning.

41.2 History of Blended Learning (BL)

The history of blended learning (BL) goes back to the 1960s when the first technology-based training approach came with mainframe and mini-computers in the 1960s and 1970s. These systems had the limitation of character-based terminals but the benefit of reaching hundreds to thousands of people at their workplace. The next step in the technology evolution came in the 1970s when companies started to use video networks to extend the live instructor. Take the problems with instructor-led training and use TV-based technology to extend the live experience. Learners could sit in a classroom, watch the instructor on TV, chat and interact with other students, and even ask the instructor questions. A well-run example of this approach is the Stanford University Interactive TV network, which is still used throughout Silicon Valley. Stanford invested in a community-based video network in the 1970s and 1980s that enables Stanford professors to teach courses all over the San Francisco Bay Area without leaving the campus. The students never have to leave their workplace to learn. The next step in the technology evolution came in the 1980s when the first PCs arrived; trainers and educators rushed head-long into PC multimedia technologies. Training technologists love to work on the cutting edge. The next step in the technology evolution came in 1998 and forwards, with First Generation Web-Based Training Virtual Classroom "E-learning." The next step in the technology evolution came in the 2002, with Integrated Blended Learning, Web, Video, and Audio, Simulations, ILT, and more. Learning Management Systems played an important role in the development of blended learning as follows:

Learning Management Systems (LMSs) are very important in blended learning in higher education as they contain an effective web-based learn-ing system of sharing study materials, making announcements, conducting evaluation and assessments, generating results, communicating interactively in synchronous and asynchronous ways, among various other academic activities (Kant et al., 2021; Bervell and Umar, 2017).

41.2.1 Moodle

Moodle is an open-source LMS that provides collaborative learning envi-ronments which empower learning and teaching. It is a flexible and user-friendly platform adopted by most educational institutions and businesses of all sizes.

41.2.2 Blackboard

Blackboard is the most popular LMS used by businesses and educational institutes worldwide, which delivers a powerful learning experience. It is easily customizable according to your organization's needs. It provides advanced features and integrates with Dropbox, Microsoft OneDrive, and school information systems.

Moodle and Blackboard are two of the most famous and widely known Learning Management Systems (LMSs) in pharmacy education and help educators as well students in the teaching and learning process (Momani, 2010).

Momani (2010) compared the two known Learning Management Systems (LMSs) in terms of pedagogical factor, learner environment, instructor tools, course and curriculum design, administrator tools, and technical specifica-tions and reported that they have lots in common, but also have some key differences which make each one special in its own way (Momani, 2010).

41.2.3 Webinar and Video Conferencing Platforms

Webinar and video conferencing platforms are very important in blended learning. Microsoft Teams, Cisco WebEx Teams, Google Meet, and Zoom are the most common webinar and video conferencing platforms in online pharmacy education (www.microsoft.com/en-us/microsoft-teams/group-chat-software; www.webex.com/; https://apps.google.com/meet/; www.zoom.us/).

Microsoft Teams, Cisco WebEx Teams, Google Meet, and Zoom offer multiple versions of their software based on usage requirements. This includes free versions that are great for light uses, short conference calls, and light file sharing. However, universities have paid the cost of all platforms for pharmacy educators and students to facilitate teaching and learning. All platforms are used successfully in online pharmacy education worldwide. Training, workshops, and writing manuals are very important for both pharmacy educators and students in order to use these platforms successfully.

41.3 Importance of Blended Learning (BL)

Blended learning (BL) has many benefits such as the following:

Flexible.

Improved access to education.

Less expensive.

Convenience.

Helps students to explore technology and use different tools or techniques for learning, for example, PowerPoint, virtual classrooms, video lectures, and so on.

Improves the quality of education and information assimilation while making teaching more efficient and productive.

Improve students' self-learning skills.

Improves students' understanding of the course materials.

Improves students' critical thinking skills.

Improves students' communication skills.

Improves students' attitude towards teamwork and collaboration.

Improves students' decision-making skills.

Improves students' time management skills.

Improves students' presentation skills.

Improves students' problem-solving skills.

Improves knowledge retention.

41.4 Purpose of Blended Learning (BL)

The aim of blended learning is to give students the opportunity to learn outside the university; to prepare students to be lifelong learners and improve their self-reading skills as well as to improve their basic and clinical knowledge and skills.

41.5 Blended Learning Models

There are many models of blended learning (blendedlearning.org).

41.5.1 Station Rotation Model

The station rotation model allows students to rotate through stations on a fixed schedule, where at least one of the stations is an online learning station.

41.5.2 Lab Rotation Model

The lab rotation model, like a station rotation, allows students to rotate through stations on a fixed schedule. However, in this case, online learning occurs in a dedicated computer lab. This model allows for flexible scheduling arrangements with teachers and other paraprofessionals, and enables schools to make use of existing computer labs.

41.5.3 Individual Rotation Model

The individual rotation model allows students to rotate through stations, but on individual schedules set by a teacher or software algorithm. Unlike other rotation models, students do not necessarily rotate to every station; they rotate only to the activities scheduled on their playlists.

41.5.4 Flipped Classroom Model

The flipped classroom model flips the traditional relationship between class time and homework. Students learn at home via online coursework and lectures, and teachers use class time for teacher-guided practice or projects. This model enables teachers to use class time for more than delivering traditional lectures.

41.5.5 Flex Model

The flex model lets students move on fluid schedules among learning activities according to their needs. Online learning is the backbone of student learning in a flex model. Teachers provide support and instruction on a flexible, as-needed basis while students work through course curriculum and content. This model can give students a high degree of control over their learning.

41.5.6 A La Carte Model

The a la carte model enables students to take an online course with an online teacher of record, in addition to other face-to-face courses, which often provides students with more flexibility over their schedules. A la carte courses can be a great option when schools can't provide particular learning opportunities, such as an advanced placement or elective course, making it one of the more popular models in blended high schools.

41.5.7 Enriched Virtual Model

The enriched virtual model is an alternative to full-time online school that allows students to complete the majority of coursework online at home or outside of school, but attend school for required face-to-face learning sessions with a teacher. Unlike the flipped classroom, enriched virtual programs usually don't require daily school attendance; some programs may only require twice-weekly attendance, for example.

41.6 Tips for Implementing Effective Blended Learning

Orientation is very important for the success of blended learning. Orientation should be about the course objectives, learning outcomes, topics, educational resources, and learning process.

Provide students with the course resources such as books, lecture notes, and others.

Facilitate communication.

Check the technology facilities at your school as well as at students' homes.

Training about technology use is very important to pharmacy educators as well as students.

Provide students with the online course resources such as e-books, lecture notes, and others.

Provide assessment and feedback.

41.7 Best Practices for Blended Learning in Pharmacy Education

Margolis et al., 2017 reported the following best practices for blended learning in pharmacy education:

Setting the Stage

Discuss blended learning on the first day of class.

Include blended learning in the course syllabus and schedule, with due dates and grading information; share estimated length of time for out-of-class activities.

If time off is given for online activities, label that time in the course schedule.

Consistency with Team Teaching

Communicate consistently with students through the same mechanism.

If different instructors use varying forms of blended learning (i.e., different technologies or activities), describe this in the syllabus.

Timeliness

Post materials at least 2 weeks prior to the due date or class.

Time on Task

Consider time compensation (i.e., cancellation of a face-to-face class) for online activities expected to take 15 minutes or longer.

Accountability

Provide course credit (e.g., completion points, quiz, assignment) for completing online materials on time.

Structured Active Learning

Focus on application of material using active learning techniques during face-to-face class time; examples include real-world patient cases, practice problems, think-pair-share, and buzz group discussions, clicker questions, and minute papers.

Faculty Feedback on Student Preparation

Incorporate student performance on pre-class activities to focus the practice and discussion during the face-to-face session.

Incorporating Student Feedback into the Course

Incorporate student suggestions into the class when feasible and appropriate; consider in-time changes to the class during the semester; report to students what changes were made based on student feedback.

Reviewing Online Material during Class

Consider a brief review of complicated topics at beginning of a face-to-face session; focus the majority of class time on active learning and application of material.

Technology

Choose technology that provides flexibility for students in completing online tasks when feasible; engage instructional and information technology (IIT) when developing and implementing blended learning and online activities.

41.8 Barriers to Blended Learning

There are many barriers to implementing blended learning as follows:

Lack of resources.
Lack of funds, financial issues.
Knowledge of pharmacy educators about blended learning.
Attitude of pharmacy educators towards blended learning.
Attitude of pharmacy students towards blended learning.
Resistance of pharmacy educators towards blended learning.
Resistance of pharmacy students towards blended learning.
University/school culture towards blended learning.
Lack of training about blended learning.
Lack of motivation to train about blended learning.
Lack of technologies to implement blended learning.

41.9 Conclusion

This chapter has discussed the history and importance of blended learning; the purpose of blended learning; models of blended learning; best practice for blended learning in pharmacy education; and barriers to blended learning implementation.

References

Learning Management Systems (LMSs)

Bersin, J., 2004. *The blended learning book: Best practices, proven methodologies, and lessons learned.* John Wiley & Sons.

Bervell, B. and Umar, I.N., 2017. A decade of LMS acceptance and adoption research in Sub-Sahara African higher education: A systematic review of models, methodologies, milestones and main challenges. *EURASIA Journal of Mathematics, Science and Technology Education*, 13(11), pp. 7269–7286.

Hrastinski, S., 2019. What do we mean by blended learning? *TechTrends*, 63(5), pp. 564–569. www.blendedlearning.org/models/

Garrison, D.R. and Kanuka, H., 2004. Blended learning: Uncovering its transformative potential in higher education. *The Internet and Higher Education*, 7(2), pp. 95–105.

Garrison, D.R. and Vaughan, N.D., 2008. *Blended learning in higher education: Framework, principles, and guidelines.* John Wiley & Sons.

Graham, C.R., 2006. *Blended learning systems: The handbook of blended learning: Global perspectives, local designs*, 1 (pp. 3–21).

Kant, N., Prasad, K.D. and Anjali, K., 2021. Selecting an appropriate learning management system in open and distance learning: A strategic approach. *Asian Association of Open Universities Journal*, 16(1), pp. 79–97.

Margolis, A.R., Porter, A.L. and Pitterle, M.E., 2017. Best practices for use of blended learning. *American Journal of Pharmaceutical Education*, 81(3).

Momani, A.M., 2010. Comparison between two learning management systems: Moodle and Blackboard. *Behavioral & Social Methods eJournal*, 2(54).

42

Massive Open Online Courses in Pharmacy Education

42.1 What Are Massive Open Online Courses (MOOCs)?

Massive open online courses (MOOCs) can be defined as follows (Maxwell et al., 2013): A massive open online course (MOOC) is a model for delivering learning content online to any person who wants to take a course, with no limit on attendance. Massive open online courses (MOOCs) are a mechanism of mass dissemination of information through an internet-based educational course to potentially very large and internationally distributed groups of learners. MOOCs engage thousands of students without geographic bounds simultaneously in an internet-based, virtual education and socialization experience. True MOOCs engage the learner actively and relationally through a variety of learning methods and media, including live chat, online learning assessments, and video, for example. MOOCs function outside the brick-and-mortar and financial model construct of higher education, allowing individuals to selectively acquire knowledge without having to enroll in a particular university, pay tuition, or commit to a degree program (Maxwell et al., 2013; Wamsley et al., 2018).

MOOC is best explained by elaborating on the four words that make up this acronym as follows (Inchiparamban and Pingle, 2016):

Massive: The word massive indicates the large student population the world over that such an online course caters to.

Open: This word, as used in MOOC, could mean any of the following:

It is open to everyone and does not require learners to fulfill certain criteria such as in terms of qualifications.

It could also indicate free access to educational resources.

Open could also refer to the fact that educational resources are hosted on open platforms such as wikis.

Lastly, it may also refer to the fact that there is use of open content extensively. This indicates that the content generated by the course is published openly for use by others.

Online: An essential feature of any MOOC is that it is online, meaning that learners can take this course using the internet. This is where it benefits learners who are located in remote locations.

Course: The word "course" indicates that a MOOC has clear learning objectives, an instructional strategy, and testing of the learning that may have occurred.

By virtue of having these features, MOOCs today provide the promise of heralding a new wave of learning in distance and adult education.

42.2 History of Massive Open Online Courses

The history of massive online open courses (MOOCs) goes back to the 2000s when the term MOOC was coined in 2008 by Dave Cormier at the University of Prince Edward Island and Bryan Alexander of the National Institute for Technology in Liberal Education in response to an open online course designed and led by George Siemens at Athabasca University and Stephen Downes at The National Research Council (Canada). The 2008 course was called "Connectivism and Connective Knowledge" and was presented to 25 tuition fee-paying students in Extended Education at the University of Manitoba in addition to 2,300 other students from the general public who took the online class free of charge. However, in 2000, two initiatives appeared, Fhatom and AllLearn, which had many of the characteristics of current initiatives such as Coursera or edX, so that they should be considered as the first MOOCs. 2007 is often observed as the year of the first MOOC, when David Wiley, professor at Utah State University, opened an official course taught to anyone who wanted to participate. In this course, 50 online students from eight countries joined the five face-to-face students.

42.3 Importance of Massive Open Online Courses (MOOCs)

Massive open online courses (MOOCs) have many benefits such as the following:

> MOOCs are getting more and more popular; this is because they offer several benefits in terms of accessibility, increased potential for student engagement, and lifelong learning opportunities. MOOCs are free or available to learners at a low cost and they are easily accessible because they are online. Being online, MOOCs attract learners because they offer a certain degree of freedom and flexibility to learners. There is no special

requirement in terms of educational background, as well. Other than that, there is no age limit to take up a MOOC and learners can access MOOCs from anywhere and at any time. Additionally, MOOCs are a great medium to ensure continuity of learning. Be it for career growth or purely for the sake of learning, MOOCs have the potential to guarantee lifelong learning. These benefits only seem to indicate that MOOCs are here to stay.

42.4 Purpose of Massive Open Online Courses (MOOCs)

The aim of massive open online courses (MOOCs) is to provide an affordable and flexible way to learn new skills.

42.5 Massive Open Online Courses (MOOCs) Models

MOOCs tend to fall into one of two models. The first is connectivist or cMOOCs, which originated in 2008 at the University of Manitoba by Stephen Downes and George Siemens, and in which online communities form around a subject of mutual interest, typically outside traditional educational contexts. The second is the xMOOC, typically offered by for-profit providers like Coursera® (www.coursera.org/) or edX® (www.edx.org/), that supply content from a central source, such as a professor (Maxwell et al., 2013).

42.6 Massive Open Online Courses (MOOCs) in Pharmacy Education

Maxwell et al. (2013) summarized the massive open online courses (MOOCs) in pharmacy education as follows (Maxwell et al., 2013):

> The Ohio State University College of Pharmacy, the OSU COP, began providing a MOOC in 2013 entitled Generation Rx: The Science Behind Prescription Drug Abuse, which was a 6-week long xMOOC. Course content was delivered through weekly introduction videos, content videos (short instructor-centered lectures with audience response opportunities), readings, discussions, quizzes, and a peer-reviewed public service announcement project. A single faculty member developed the course material with the assistance of media experts and OSU's Office of Distance Education and eLearning.

In 2013, a team at The University of Texas at Austin College of Pharmacy created, developed, and implemented an xMOOC called Take Your Medicine (TYM) using the edX® platform. TYM was developed as an 8-week course to demystify the drug development process, explore how research innovations are turned into medications, and teach how to be a savvy consumer and patient. The initial course was divided into two 4-week mini-MOOCs, "Developing New Drug Products" and "How to be a Savvy Consumer." Interactive dialogues and gaming prototypes were incorporated using Google's Oppia tool and Articulate Storyline©. These elements enabled students to make choices about the development of new drug products, participate in interaction dialogues, and presented the learner with feedback on their choices.

42.7 Massive Open Online Courses (MOOCs) Challenges

There are many challenges for massive open online courses as follows (Wikipedia, 2021):

Relying on user-generated content can create a chaotic learning environment.

Digital literacy is necessary to make use of the online materials.

The time and effort required from participants may exceed what students are willing to commit to a free online course.

Once the course is released, content will be reshaped and reinterpreted by the massive student body, making the course trajectory difficult for instructors to control.

Participants must self-regulate and set their own goals.

Language and translation barriers.

Accessibility barriers for differently abled users.

Access barriers for people from low socioeconomic neighborhoods and countries with very little internet access.

42.8 Conclusion

This chapter has discussed the history and importance of massive open online courses (MOOCs); the purpose of MOOCs; MOOCs and challenges of MOOCs.

References

Inchiparamban, S. and Pingle, S., 2016. Massive open online courses (MOOCs): Why do we need them? *Online Submission*. Available at: https://files.eric.ed.gov/fulltext/ED590312.pdf

Maxwell, W.D., Fabel, P.H., Diaz, V., Walkow, J.C., Kwiek, N.C., Kanchanaraksa, S. and Waldrop, M.M., 2013. Online learning: Campus 2.0. *Nature News*, 495(7440), p. 160.

Wamsley, M., Chen, A. and Bookstaver, P.B., 2018. Massive open online courses in US healthcare education: Practical considerations and lessons learned from implementation. *Currents in Pharmacy Teaching and Learning*, 10(6), pp. 736–743.

Wikipedia., 2021. *Massive open online course*. Available at: https://en.wikipedia.org/wiki/Massive_open_online_course

43

Computer-Assisted Learning and Computer-Based Learning in Pharmacy Education

43.1 What Is Computer Assisted Learning (CAL) & Computer-Based Learning (CBL)?

Computer-assisted learning (CAL) can be defined as follows (Ranga et al., 2017; Nerlich, 1995; Schittek et al., 2001): any learning that is mediated by a computer and that requires no direct interaction between the user and a human instructor in order to run. Instead, CAL presents the user with an interface (constructed by an educator skilled in the field of study) which allows the user to follow a lesson plan or may allow self-directed access to particular information of interest. Computer-assisted learning (CAL) can be defined as the use of instructional tools presented and managed by a computer. Instructional computers either present information or fill a tutorial role, testing the student for comprehension, giving the student feedback for overcoming difficulties, and guiding the student in recovery actions when errors and/or mistakes appear. There are many different names used for this, including computer-based teaching (CBT) and computer-assisted instruction (CAI) (Ranga et al., 2007; Nerlich, 1995).

Computer-based learning (CBL) can be defined as follows (Rogers et al., 2009):

Use of a computer to deliver instructions to students using a variety of instructional strategies to meet individual students' needs.

Refers to the use of computers as a key component of the educational environment. Broadly refers to a structured environment in which computers are used for teaching purposes.

Computer-based learning (CBL) is the term used for any kind of learning with the help of computers.

CBL refers to any kind of learning that involves the use of the interactive elements of computer applications and software.

DOI: 10.1201/9781003230458-47

43.2 History of Computer-Assisted Learning (CAL) and Computer-Based Learning (CBL)

The history of computer-assisted learning (CAL) and computer-based learning (CBL) goes back to the 1950s; by 1966, however, IBM had developed its 1500 CAI System for research and development of CAI lessons (Blaisdell, 1978).

43.3 Purpose of Computer-Assisted Learning (CAL) and Computer-Based Learning (CBL)

The aim of computer-assisted learning (CAL) is to enhance the teaching and learning process by using computer-related technologies. In computer-based learning (CBL), the computer is used for instructional purposes where the computer hardware and software as well as the peripherals and input devices are key components of the educational environment. CBL assists individuals in learning using multiple representations of information for a specific educational purpose. Common innovative realizations of CBL to improve teaching and learning are hypertext, simulations, and microworlds.

43.4 Importance of Computer-Assisted Learning (CAL) and Computer-Based Learning (CBL)

Computer-assisted learning (CAL) and computer-based learning (CBL) have many benefits such as the following:

Improves students' understanding of the course materials.

Improves students' critical thinking skills.

Improves students' communication skills.

Improves students' attitude towards teamwork and collaboration.

Improves students' self-learning skills.

Improves students' decision-making skills.

Improves students' time management skills.

Improves students' presentation skills.

Improves students' problem-solving skills.

Improves knowledge retention.

Improves leadership skills.

Improves the ability to answer the cases.

Improves the ability to provide/design pharmacist care and patient care.

Improves the ability to identify and solve/prevent drug-related problems (DRPs).

43.5 Applications of Computer-Assisted Learning (CAL) and Computer-Based Learning (CBL) in Pharmacy Education

Computer-assisted learning has many applications in pharmacy education such as:

Teaching theory.

Teaching tutorials.

Training.

Simulation.

Web-based education.

Evaluations and assessments.

Research.

Literature review tool.

Storing electronic books and using them easily.

Tool for online teaching and learning.

43.5.1 Online Computer-Based Learning (CBL)

Computer-based learning (CBL) can be implemented online with the help of new technologies and facilities such as:

43.5.2 Internet

The internet plays a very important and vital role in online pharmacy education. The internet facilitates the teaching and learning process which makes distance/online pharmacy education more effective and easier than at any time in history. Pharmacy educators and students need the internet for: making communication easy; delivering the classes; uploading/

downloading the lecture notes and other educational materials/resources; exams; assignments; presentations; searching for pharmacy-related information; training; patient care services; public health promotion, awareness, and services. Without access to the internet, online teaching and learning will stop.

43.5.3 Computers and Laptops

Using computers and laptops is very important and essential for online teaching and learning. It provides flexible and effective access to online teaching and learning for pharmacy educators and students.

43.5.4 Learning Management Systems (LMSs)

Learning Management Systems (LMSs) are very important in online pharmacy education as well as higher education, as it contains an effective web-based learning system of sharing study materials, making announcements, conducting evaluation and assessments, generating results, communicating interactively in synchronous and asynchronous ways, among various other academic activities (Kant et al., 2021; Bervell and Umar, 2017).

43.5.5 Moodle

Moodle is an open-source LMS that provides collaborative learning environments which empower learning and teaching. It is a flexible and user-friendly platform adopted by most educational institutions and businesses of all sizes.

43.5.6 Blackboard

Blackboard is the most popular LMS used by businesses and educational institutes worldwide, which delivers a powerful learning experience. It is easily customizable according to your organization's needs. It provides advanced features and integrates with Dropbox, Microsoft OneDrive, and school information systems.

Moodle and Blackboard are two of the most famous and widely known learning management systems (LMSs) in pharmacy education and help educators as well students in the teaching and learning process (Momani, 2010).

Momani (2010) compares the two known learning management systems (LMSs) in terms of pedagogical factor, learner environment, instructor tools, course and curriculum design, administrator tools and technical specifications and reported that they have lots in common, but also have some key differences which make each one special in its own way (Momani, 2010).

43.5.7 Webinar and Video Conferencing Platforms

Webinar and video conferencing platforms are very important in online pharmacy education. Microsoft Teams, Cisco WebEx Teams, Google Meet, and Zoom are the most common webinar and video conferencing platforms in online pharmacy education (www.microsoft.com/en-us/microsoft-teams/group-chat-software; www.webex.com/; https://apps.google.com/meet/; www.zoom.us/).

Microsoft Teams, Cisco WebEx Teams, Google Meet, and Zoom offer multiple versions of their software based on usage requirements. This includes free versions that are great for light uses, short conference calls, and light file sharing. However, universities have paid the cost of all platforms for pharmacy educators and students to facilitate teaching and learning. All platforms are used successfully in online pharmacy education worldwide. Training, workshops, and writing manuals are very important for both pharmacy educators and students in order to use these platforms successfully.

43.6 Online and Digital Library

The online and digital library is a collection of documents such as journal articles, books, and other educational resources organized in an electronic form and available on the internet for students, educators, and other staff. Furthermore, access is provided to the databases which help students, educators, and staff to access the latest volumes/issues of scientific journals.

43.7 Barriers to Computer-Based Learning (CBL)

Lack of resources.

Lack of funds, financial issues.

Knowledge of pharmacy educators about computer-based learning (CBL).

Attitude of pharmacy educators towards CBL.

Attitude of pharmacy students towards CBL.

Resistance of pharmacy educators towards CBL.

Resistance of pharmacy students towards CBL.

University/school culture towards CBL.

Lack of training about CBL.

Lack of motivation to train about CBL.

Lack of technologies to implement CBL.

43.8 Conclusion

This chapter has discussed the history and importance of computer-assisted Learning (CAL); the purpose of computer-assisted learning (CAL); the applications of computer-assisted learning (CAL) and computer-based learning (CBL) in pharmacy education.

References

Bervell, B. and Umar, I.N., 2017. A decade of LMS acceptance and adoption research in Sub-Sahara African higher education: A systematic review of models, methodologies, milestones and main challenges. *EURASIA Journal of Mathematics, Science and Technology Education*, 13(11), pp. 7269–7286.

Blaisdell, F.J., 1978. Historical development of computer assisted instruction. *Journal of Technical Writing and Communication*, 8(3), pp. 253–268.

Kant, N., Prasad, K.D. and Anjali, K., 2021. Selecting an appropriate learning management system in open and distance learning: A strategic approach. *Asian Association of Open Universities Journal*, 16(1), pp. 79–97.

Momani, A.M., 2010. Comparison between two learning management systems: Moodle and blackboard. *Behavioral & Social Methods eJournal*, 2(54).

Nerlich, S., 1995. Computer-assisted learning (CAL) for general and specialist nursing education. *Australian Critical Care*, 8(3), pp. 23–27.

Ranga, V., Koul, B.N. and Thomas, K., 2017. Unit-1 computer as an educational aid. Rogers, P.L., Berg, G.A., Boettcher, J.V., Howard, C., Justice, L. and Schenk, K.D. eds. 2009. *Encyclopedia of distance learning*. IGI Global.

Schittek, M., Mattheos, N., Lyon, H.C. and Attström, R., 2001. Computer assisted learning. A review. *European Journal of Dental Education: Review Article*, 5(3), pp. 93–100.

Section 5

Pharmacy Education Assessment and Evaluation Methods

44

Assessment Methods in Pharmacy Education: Strengths and Limitations

44.1 Terminologies

44.1.1 Assessment and Evaluation

Mavis (2014) differentiates between assessment and evaluation as follows: assessment most often refers to the measurement of individual student performance, while evaluation refers to the measurement of outcomes for courses, educational programs, or institutions. Practically speaking, students are assessed while educational programs are evaluated. However, it is often the case that aggregated student assessments serve as an important information source when evaluating educational programs (Mavis, 2014).

44.1.2 Formative and Summative Assessment

Literature reports the following definitions and differentiates between formative and summative assessment (Black and Wiliam, 2010; Garrison and Ehringhaus, 2007; Dixson and Worrell, 2016; Mavis, 2014). Formative assessment has been defined as "activities undertaken by teachers—and by their students in assessing themselves—that provide information to be used as feedback to modify teaching and learning activities." The purpose of formative assessment is to monitor student learning and provide ongoing feedback to staff and students. It is assessment for learning. If designed appropriately, it helps students identify their strengths and weaknesses, and enables students to improve their self-regulatory skills so that they manage their education in a less haphazard fashion than is commonly found. It also provides information to the faculty about the areas students are struggling with so that sufficient support can be put in place. Formative assessment can be tutor led, peer or self-assessment. Formative assessments have low stakes and usually carry no grade, which in some instances may discourage students from doing the task or fully engaging with it (Black and Wiliam, 2010; Garrison and Ehringhaus, 2007; Dixson and Worrell, 2016; Mavis, 2014).

DOI: 10.1201/9781003230458-49

Summative assessments are given periodically to determine at a particular point in time what students know and do not know. Many associate summative assessments only with standardized tests such as state assessments, but they are also used in and are an important part of district and classroom programs. Summative assessment at the district/classroom level is an accountability measure that is generally used as part of the grading process (Black and Wiliam, 2010; Garrison and Ehringhaus, 2007; Dixson and Worrell, 2016; Mavis, 2014).

44.1.3 Diagnostic Assessment

Diagnostic assessment is used to give information about students' prior knowledge. It helps in designing the most suitable educational program for each student (Garrison and Ehringhaus, 2007; Alfadl, 2018).

44.1.4 Direct Assessment

Direct assessment refers to assessment that is based on an analysis of student behaviors or products in which they demonstrate how well they have mastered learning outcomes (Allen, 2004).

44.1.5 Indirect Assessment

Indirect assessment refers to assessment that is based on gathering information through means other than looking at actual samples of student work such as surveys, interviews, and focus groups. Indirect assessment refers to any method of collecting data that requires reflection on student learning, skills, or behaviors, rather than a demonstration of it; it is indirect evidence of student achievement which requires that faculty infer actual student abilities, knowledge, and values rather than observing direct evidence of learning or achievement.

44.2 Rationality of Assessment and Evaluation in Pharmacy Education

Assessment and evaluation in pharmacy education at the level of programs, curriculum, and courses are very important to assess the quality of pharmacy education, to ensure that the graduates are competent, have the required competencies, and are able to provide good pharmacist care and patient care services. Literature reports the following reasons for assessing students' performance (Mavis, 2014; McAleer, 2001):

- Providing feedback to students about their mastery of course content.
- Grading or ranking students for progress and promotion decisions.
- Offering encouragement and support to students (or teachers).
- Measuring changes in knowledge, skills, or attitudes over time.
- Diagnosing weaknesses in student performance.
- Establishing performance expectations for students.
- Identifying areas for improving instruction.
- Documenting instructional outcomes for faculty promotion.
- Evaluating the extent to which educational objectives are realized.
- Encouraging the development of a new curriculum.
- Demonstrating quality standards for the public, institution, or profession.
- Articulating the values and priorities of the educational institution.
- Informing the allocation of educational resources.
- It is an integral part of the learning process in which students are informed of any weaknesses and of how to improve on the quality of their performance.
- It illustrates progress and ensures a proper standard has been achieved before progressing to a higher level of training.
- It provides certification relating to a standard of performance, for example, the award of a degree.
- It indicates to students the areas of a course which are considered important.
- It acts as a promotion technique.
- It acts as a means of selection for a career or as an entrance requirement for a course.
- It motivates students in their studies.
- It measures the effectiveness of training and identifies curriculum weaknesses.

44.3 Key Features of Student Assessment Methods

Assessment methods should have the following (Mavis, 2014):

1. Reliability
2. Validity

3. Feasibility
4. Acceptability
5. Educational Impact

44.4 Strengths and Limitations of Assessment and Evaluation in Pharmacy Education

Each assessment and evaluation method either, summative or formative, has strengths/advantages and limitations/weakness (Morningside College, 2006).

44.4.1 Standardized Exams

Advantages

- Convenient.
- Can be adopted and implemented quickly.
- Reduces or eliminates faculty time demands in instrument development and grading.
- Are scored objectively.
- Provide for external validity.
- Provide reference group measures.
- Can make longitudinal comparisons.
- Can test large numbers of students.

Disadvantages

- Measures relatively superficial knowledge or learning.
- Unlikely to match the specific goals and objectives of a program/institution.
- Norm-referenced data may be less useful than criterion-referenced.
- May be cost prohibitive to administer as a pre- and post-test.
- More summative than formative (may be difficult to isolate what changes are needed).
- Norm data may be user norms rather than true national sample.
- May be difficult to receive results in a timely manner.

44.4.2 Performance Measures

Types

- Essays
- Oral presentations
- Oral exams
- Exhibitions
- Demonstrations
- Performances
- Products
- Research papers
- Poster presentations
- Capstone experiences
- Practical exams
- Supervised internships and practicums

Advantages

- Can be used to assess from multiple perspectives.
- Using a student-centered design can promote student motivation.
- Can be used to assess transfer of skills and integration of content.
- Engages student in active learning.
- Encourages time on academics outside of class.
- Can provide a dimension of depth not available in the classroom.
- Can promote student creativity.
- Can be scored holistically or analytically.
- May allow probes by faculty to gain a clearer picture of student understanding or thought processes.
- Can provide closing of feedback loop between students and faculty.
- Can place faculty more in a mentor role than as judge.
- Can be summative or formative.
- Can provide an avenue for student self-assessment and reflection.
- Can be embedded within courses.
- Can adapt current assignments.
- Usually the most valid way of assessing skill development.

Disadvantages

- Usually the mostly costly approach.
- Time-consuming and labor intensive to design and execute for faculty and students.
- Must be carefully designed if used to document obtainment of student learning outcomes.
- Ratings can be more subjective.
- Requires careful training of raters.
- Inter-rater reliability must be addressed.
- Production costs may be prohibitive for some students and hamper reliability.
- Sample of behavior or performance may not be typical, especially if observers are present.

44.4.3 Portfolios

Potential Advantages

- Shows sophistication in student performance.
- Illustrates longitudinal trends.
- Highlights student strengths.
- Identifies student weaknesses for remediation, if timed properly.
- Can be used to view learning and development longitudinally.
- Multiple components of the curriculum can be assessed (e.g., writing, critical thinking, technology skills).
- Samples are more likely than test results to reflect student ability when planning, input from others, and similar opportunities common to more work settings are available.
- Process of reviewing and evaluating portfolios provides an excellent opportunity for faculty exchange and development, discussion of curriculum goals and objectives, review of criteria, and program feedback.
- May be economical in terms of student time and effort if no separate assessment administration time is required.
- Greater faculty control over interpretation and use of results.
- Results are more likely to be meaningful at all levels (student, class, program, institution) and can be used for diagnostic and prescriptive purposes as well.
- Avoids or minimizes test anxiety and other one-shot measurement problems.
- Increases power of maximum performance measures over more artificial or restrictive speed measures on test or in-class samples.

- Increases student participation (selection, revision, and evaluation) in the assessment process.
- Could match well with Morningside's mission to cultivate lifelong learning.
- Can be used to gather information about students' assignments and experiences.
- Reflective statements could be used to gather information about student satisfaction.

Potential Disadvantages

- Portfolio will be no better than the quality of the collected artifacts.
- Time-consuming and challenging to evaluate.
- Space and ownership challenges make evaluation difficult.
- Content may vary widely among students.
- Students may fail to remember to collect items.
- Transfer students may not be in the position to provide a complete portfolio.
- Time intensive to convert to meaningful data.
- Costly in terms of evaluator time and effort.
- Management of the collection and evaluation process, including the establishment of reliable and valid grading criteria, is likely to be challenging.
- May not provide for externality.
- If samples to be included have been previously submitted for course grades, faculty may be concerned that a hidden agenda of the process is to validate their grading.
- Security concerns may arise as to whether submitted samples are the students' own work or adhere to other measurement criteria.
- Must consider whether and how graduates will be allowed continued access to their portfolios.
- Inter-rater reliability must be addressed.

44.5 Selection of Assessment and Evaluation Tools in Pharmacy Education

Literature reports the following steps for developing the good assessment instrument (Hamstra, 2014; Hamstra, 2012):

1. Determine the purpose of your assessment.

   A. Formative, summative (standard setting/criteria) research

   B. Knowledge, skills, attitudes (e.g., performance, teamwork, anxiety)

2. Content validity—identify the main construct of interest and stakeholders.

3. Review with content experts—focus group.

   A. Representative sample: different institutions and disciplines

   B. Thematic saturation, address political issues

   C. Set preliminary standards—what does perfect/borderline performance look like?

4. Item writing/development (based on related existing tests?)

5. If necessary, train the raters (and assess inter-rater reliability).

6. Pilot test the instrument (representative sample) for validity.

   A. Feasibility check—length, clarity, cost

   B. If necessary, go back to Step 4 (modify items and pilot test again).

7. Implement modified test—measure reliability, validity based on larger sample.

   A. Assess construct validity.

44.6 Conclusion

This chapter has discussed the types of assessment and evaluations; the rationality of assessment and evaluation in pharmacy education; key features of student assessment methods and the selection of assessment and evaluation tools in pharmacy education. Selecting the appropriate assessment and evaluation methods is very important to assess and evaluate the students' performance. Furthermore, assessment and evaluation methods are very important to evaluate the achievement of program learning outcomes as well as the course learning outcomes.

References

Alfadl, A.A., 2018. Assessment methods and tools for pharmacy education. In *Pharmacy education in the twenty first century and beyond* (pp. 147–168). Academic Press.

Allen, M.J., 2004. *Assessing academic programs in higher education* (p. 2013). Anker Publishing. Retrieved February, 6.

Black, P. and Wiliam, D., 2010. Inside the black box: Raising standards through classroom assessment. *Phi Delta Kappan*, 92(1), pp. 81–90.

Dixson, D.D. and Worrell, F.C., 2016. Formative and summative assessment in the classroom. *Theory into Practice*, 55(2), pp. 153–159.

Garrison, C. and Ehringhaus, M., 2007. *Formative and summative assessments in the classroom*. Available at: http://ccti.colfinder.org/sites/default/files/formative_and_summative_assessment_in_the_classroom.pdf

Hamstra, S.J., 2012. Keynote address: The focus on competencies and individual learner assessment as emerging themes in medical education research. *Academic Emergency Medicine*, 19(12), pp. 1336–1343.

Hamstra, S.J., 2014. *Designing and selecting assessment instruments: Focusing on competencies. The royal college program directors handbook: A practical guide for leading an exceptional program*. Royal College of Physicians and Surgeons of Canada.

Mavis, B., 2014. Assessing student performance. In *An introduction to medical teaching* (pp. 209–241). Springer.

McAleer, S., 2001. *Choosing assessment instruments: A practical guide for medical teachers* (p. 303313). Churchill Livingstone.

Morningside College., 2006. *Assessment handbook. Advantages and disadvantages of various assessment methods*. Available at: https://www.clark.edu/tlc/outcome_assessment/documents/AssessMethods.pdf

45

Assessment Methods in Pharmacy Education: Direct Assessment

45.1 Background

Direct assessment refers to assessment that is based on an analysis of student behaviors or products in which they demonstrate how well they have mastered learning outcomes (Allen, 2004). Good assessment should be (Allen, 2004):

- Valid—directly reflects the learning outcome being assessed.
- Reliable—especially inter-rater reliability when subjective judgments are made.
- Actionable—results help faculty identify what students are learning well and what requires more attention.
- Efficient and cost-effective in time and money.
- Engaging to students and other respondents—so they'll demonstrate the extent of their learning.
- Interesting to faculty and other stakeholders—they care about results and are willing to act on them.
- Triangulation—multiple lines of evidence point to the same conclusion.

45.2 Direct Assessment Methods and Their Application in Pharmacy Education

Literature reported the following examples of direct assessment methods (Allen, 2004).

45.2.1 Published/Standardized Tests

Published tests or standardized tests are instruments that have been commercially published by a test publisher. These instruments are administered and scored in a consistent, or "standard" manner. The validity and reliability of the instrument are two essential elements for defining the standard quality of the test. These tests are generally only available from the publisher and often come in the form of kits or multiple booklets. They can be very costly if purchased.

Examples of Published/Standardized Tests in Pharmacy Education

Pharmacotherapy books, contain questions about each topic.
Pharmacokinetics books, contain questions about each topic.
Clinical pharmacokinetics books, contain questions about each topic.
Pharmacology books, contain questions about each topic.
Others.

Steps in Selecting a Published Test

1. Identify a possible test.
2. Consider published reviews of this test.
3. Order a specimen set from the publisher.
4. Take the test and consider the appropriateness of its format and content.
5. Consider the test's relationship to your learning outcomes.
6. Consider the depth of processing of the items (e.g., analyze items using Bloom's taxonomy).
7. Consider the publication date and currency of the items.
8. How many scores are provided? Will these scores be useful? How?
9. Look at the test manual. Were test development procedures reasonable? What is the evidence for the test's reliability and validity for the intended use?
10. If you will be using the norms, consider their relevance for your purpose.
11. Consider practicalities, for example, timing, test proctoring, and test scoring requirements.
12. Verify that faculty are willing to act on the results.

Strengths of Published/Standardized Tests

- Can provide direct evidence of student mastery of learning outcomes.
- They generally are carefully developed, highly reliable, professionally scored, and nationally normed.
- They frequently provide a number of norm groups, such as norms for community colleges, liberal arts colleges, and comprehensive universities.
- Online versions of tests are increasingly available, and some provide immediate scoring.
- Some publishers allow faculty to supplement tests with their own items, so tests can be adapted to better serve local needs.

Weaknesses of Published/Standardized Tests

- Students may not take the test seriously if test results have no impact on their lives.
- These tests are not useful as direct measures for program assessment if they do not align with local curricula and learning outcomes.
- Test scores may reflect criteria that are too broad for meaningful assessment.
- Most published tests rely heavily on multiple-choice items which often focus on specific facts, but program learning outcomes more often emphasize higher-level skills.
- If the test does not reflect the learning outcomes that faculty value and the curricula that students experience, results are likely to be discounted and inconsequential.
- Tests can be expensive.
- The marginal gain from annual testing may be low.
- Faculty may object to standardized exam scores on general principles, leading them to ignore results.

45.2.2 Locally Developed Tests

Faculty may decide to develop their own internal tests that reflect the program's learning outcomes.

Examples of Locally Developed Tests in Pharmacy Education

Multiple-choice questions (MCQs) and other questions developed by faculty in:

Pharmacotherapy modules.

Pharmacokinetics and clinical pharmacokinetics courses.

Pharmacology courses.

Others.

Strengths of Locally Developed Tests

- Can provide direct evidence of student mastery of learning outcomes.
- Appropriate mixes of essay and objective questions allow faculty to address various types of learning outcomes.
- Students generally are motivated to display the extent of their learning if they are being graded on the work.
- If well-constructed, they are likely to have good validity.
- Because local faculty write the exam, they are likely to be interested in results and willing to use them.
- Can be integrated into routine faculty workloads.
- The evaluation process should directly lead faculty into discussions of student learning, curriculum, pedagogy, and student support services.

Weaknesses of Locally Developed Tests

- These exams are likely to be less reliable than published exams.
- Reliability and validity generally are unknown.
- Creating and scoring exams takes time.
- Traditional testing methods have been criticized for not being "authentic."
- Norms generally are not available.

45.2.3 Embedded Assignments and Course Activities

Embedded assignments and course activities are assignments, activities, or exercises that are done as part of a class, but that are used to provide assessment data about a particular learning outcome.

Examples of Embedded Assignments and Course Activities in Pharmacy Education

- Community-service learning and other fieldwork activities such as awareness programs in the malls, others.
- Culminating projects, such as papers in capstone courses.

- Exams or parts of exams.
- Group projects.
- Homework assignments.
- In-class presentations.
- Student recitals and exhibitions.
- Comprehensive exams, theses, dissertations, and defense interviews.

Strengths of Embedded Assignments and Course Activities

- Can provide direct evidence of student mastery of learning outcomes.
- Out-of-class assignments are not restricted to time constraints typical for exams.
- Students are generally motivated to demonstrate the extent of their learning if they are being graded.
- Can provide authentic assessment of learning outcomes.
- Can involve CSL or other fieldwork activities and ratings by fieldwork supervisors.
- Can provide a context for assessing communication and teamwork skills.
- Can be used for grading as well as assessment.
- Faculty who develop the procedures are likely to be interested in results and willing to use them.
- The evaluation process should directly lead faculty into discussions of student learning, curriculum, pedagogy, and student support services.
- Data collection is unobtrusive to students.

Weaknesses of Embedded Assignments and Course Activities

- Requires time to develop and coordinate.
- Requires faculty trust that the program will be assessed, not individual teachers.
- Reliability and validity generally are unknown.
- Norms generally are not available.

45.2.4 Portfolios

Medical student portfolios are used in a variety of ways to help institutions and students assess and track learner progress. In its original definition, a portfolio is a collection of drawings or papers that represent a compilation of a person's work. With the transition to competency-based assessments and the use of frameworks such as milestones and entrustable professional activities (EPAs), portfolios have become a valuable tool in many medical schools. Student portfolios can assist institutions to meaningfully display assessment evidence, enabling longitudinal tracking and documentation of student achievement. With the appropriate supporting processes, portfolios can foster student skills in self-assessment and ongoing professional development (Dallaghan et al., 2020). A portfolio can be generally defined as: "a purposeful collection of student work that exhibits the student's efforts, progress, and achievements in one or more areas. The collection must include student participation in selecting contents, the criteria for selection, the criteria for judging merit, and evidence of student self-reflection" (Paulson et al., 1991).

Examples of Portfolios in Pharmacy Education

Strengths of Portfolios

- Can provide direct evidence of student mastery of learning outcomes.
- Students are encouraged to take responsibility for and pride in their learning.
- Students may become more aware of their own academic growth.
- Can be used for developmental assessment and can be integrated into the advising process to individualize student planning.
- Can help faculty identify curriculum gaps, lack of alignment with outcomes.
- Students can use portfolios and the portfolio process to prepare for graduate school or career applications.
- The evaluation process should directly lead faculty into discussions of student learning, curriculum, pedagogy, and student support services.
- E-portfolios or DVDs can be easily viewed, duplicated, and stored.

Weaknesses of Portfolios

- Requires faculty time to prepare the portfolio assignment and assist students as they prepare them.
- Requires faculty analysis and, if graded, faculty time to assign grades.
- May be difficult to motivate students to take the task seriously.
- May be more difficult for transfer students to assemble the portfolio if they haven't saved relevant materials.
- Students may refrain from criticizing the program if their portfolio is graded or if their names will be associated with portfolios during the review.

45.3 Best Practices in Selecting Direct Assessment Methods in Pharmacy Education

To select and apply the most valid and reliable direct assessment methods, pharmacy educators should take into consideration the following:

Select the appropriate direct assessment methods based on the literature and experience, and the method should be valid and reliable as well as approved by the school curriculum committee.

If you need to implement a new assessment method, take the approval from the department and curriculum committee.

Distribute the assessment methods based on the course competencies and course learning outcomes as follows: select and apply the appropriate assessment methods for each course learning outcome; select and apply the appropriate assessment methods for each week's lectures, labs, and tutorials from week 1 to week 16 as an example.

Allocate marks for each assessment method, for each week, for each learning outcome based on the size of the learning outcome and other factors.

Revise the assessment methods on a yearly basis and modify them if needed.

45.4 Conclusion

This chapter has discussed direct assessment methods in pharmacy education. This chapter includes background about direct assessment methods. It also includes direct assessment methods and their application in pharmacy education such as published/standardized tests and examples about it in pharmacy education, steps for selecting it, as well as the strengths and limitations of this type of assessment. Furthermore, it describes the best practices in selecting direct assessment methods in pharmacy education.

References

Allen, M.J., 2004. *Assessing academic programs in higher education* (p. 2013). Anker Publishing. Retrieved February, 6.

Dallaghan, G.L.B., Coplit, L., Cutrer, W.B. and Crow, S., 2020. Medical student portfolios: their value and what you need for successful implementation. *Academic Medicine*, 95(9), p. 1457.

Paulson, F.L., Paulson, P.R. and Meyer, C.A., 1991. What makes a portfolio a portfolio. *Educational Leadership*, 48(5).

46

Assessment Methods in Pharmacy Education: Indirect Assessment

46.1 Background

Indirect assessment refers to assessment based on gathering information through means other than looking at actual samples of student work such as surveys, interviews, and focus groups. Indirect assessment refers to any method of collecting data that requires reflection on student learning, skills, or behaviors, rather than a demonstration of it; it is indirect evidence of student achievement that requires that faculty infer actual student abilities, knowledge, and values rather than observing direct evidence of learning or achievement.

A good indirect assessment should be (Allen, 2004):

- Valid—directly reflects the learning outcome being assessed.
- Reliable—especially inter-rater reliability when subjective judgments are made.
- Actionable—results help faculty identify what students are learning well and what requires more attention.
- Efficient and cost-effective in time and money.
- Engaging to students and other respondents—so they'll demonstrate the extent of their learning.
- Interesting to faculty and other stakeholders—they care about results and are willing to act on them.
- Triangulation—multiple lines of evidence point to the same conclusion.

46.2 Indirect Assessment Methods and Their Application in Pharmacy Education

Literature reports the following examples of indirect assessment methods (Allen, 2004).

46.2.1 Surveys

Survey is an examination of opinions, behavior, and so on, made by asking people questions (Cambridge Dictionary).

Point-of-contact surveys.

Online, e-mailed, registration, or grad check surveys.

Keep it simple!

Examples of Surveys in Pharmacy Education

Surveys and questionnaires are given to stakeholders to ask them about the graduates.

Surveys and questionnaires to students, alumni, employers, and the public about any issue.

Survey Types

Checklist

Frequency

Likert scale

Linear rating scale

Others

Strengths of Surveys

- Are flexible in format and can include questions about many issues.
- Can be administered to large groups of respondents.
- Can easily assess the views of various stakeholders.
- Usually has face validity—the questions generally have a clear relationship to the outcomes being assessed.
- Tend to be inexpensive to administer.
- Can be conducted relatively quickly.
- Responses to close-ended questions are easy to tabulate and to report in tables or graphs.

- Open-ended questions allow faculty to uncover unanticipated results.
- Can be used to track opinions across time to explore trends.
- Are amenable to different formats, such as paper-and-pencil or online formats.
- Can be used to collect opinions.

Weaknesses of Surveys

- Provides indirect evidence about student learning.
- Their validity depends on the quality of the questions and response options.
- Conclusions can be inaccurate if biased samples are obtained.
- Results might not include the full array of opinions if the sample is small.
- What people say they do or know may be inconsistent with what they actually do or know.
- Open-ended responses can be difficult and time-consuming to analyze.

46.2.2 Interviews

- Interviews can be conducted one-on-one, in small groups, or over the phone.
- Interviews can be structured (with specified questions) or unstructured (a more open process).
- Questions can be close-ended (e.g., multiple-choice style) or open-ended (respondents construct a response).
- Can target students, graduating seniors, alumni, employers, community members, faculty, and so on.
- Can do exit interviews or pre-post interviews.
- Can focus on student experiences, concerns, or attitudes related to the program being assessed.
- Generally should be conducted by neutral parties to avoid bias and conflict of interest.

Tips for Effective Interviews

- Conduct the interview in an environment that allows the interaction to be confidential and uninterrupted.
- Demonstrate respect for the respondents as participants in the assessment process rather than as subjects. Explain the purpose of

the project, how the data will be used, how the respondents' anonymity or confidentiality will be maintained, and the respondents' rights as participants. Ask if they have any questions.

- Put the respondents at ease. Do more listening than talking. Allow respondents to finish their statements without interruption.

- Match follow-up questions to the project's objectives. For example, if the objective is to obtain student feedback about student advising, don't spend time pursuing other topics.

- Do not argue with the respondent's point of view, even if you are convinced that the viewpoint is incorrect. Your role is to obtain the respondents' opinions, not to convert them to your perspective.

- Allow respondents time to process the question. They may not have thought about the issue before, and they may require time to develop a thoughtful response.

- Paraphrase to verify that you have understood the respondent's comments. Respondents will sometimes realize that what they said isn't what they meant, or you may have misunderstood them. Paraphrasing provides an opportunity to improve the accuracy of the data.

- Make sure you know how to record the data and include a backup system. You may be using a tape recorder—if so, consider supplementing the tape with written notes in case the recorder fails or the tape is faulty. Always build in a system for verifying that the tape is functioning or that other data recording procedures are working. Don't forget your pencil and paper!

Strengths of Interviews

- Are flexible in format and can include questions about many issues.
- Can assess the views of various stakeholders.
- Usually has face validity—the questions generally have a clear relationship to the outcomes being assessed.
- Can provide insights into the reasons for participants' beliefs, attitudes, and experiences.
- Interviewers can prompt respondents to provide more detailed responses.
- Interviewers can respond to questions and clarify misunderstandings.
- Telephone interviews can be used to reach distant respondents.
- Can provide a sense of immediacy and personal attention for respondents.
- Open-ended questions allow faculty to uncover unanticipated results.

Weaknesses of Interviews

- Generally provides indirect evidence about student learning.
- Their validity depends on the quality of the questions.
- Poor interviewer skills can generate limited or useless information.
- Can be difficult to obtain a representative sample of respondents.
- What people say they do or know may be inconsistent with what they actually do or know.
- Can be relatively time-consuming and expensive to conduct, especially if interviewers and interviewees are paid or if the no-show rate for scheduled interviews is high.
- The process can intimidate some respondents, especially if asked about sensitive information and their identity is known to the interviewer.
- Results can be difficult and time-consuming to analyze.
- Transcriptions of interviews can be time-consuming and costly.

46.2.3 Focus Groups

- Traditional focus groups are free-flowing discussions among small, homogeneous groups (typically from 6 to 10 participants), guided by a skilled facilitator who subtly directs the discussion in accordance with predetermined objectives. This process leads to in-depth responses to questions, generally with full participation from all group members. The facilitator departs from the script to follow promising leads that arise during the interaction.
- Structured group interviews are less interactive than traditional focus groups and can be facilitated by people with less training in group dynamics and traditional focus group methodology. The group interview is highly structured, and the report generally provides a few core findings, rather than an in-depth analysis.

Strengths of Focus Groups

- Are flexible in format and can include questions about many issues.
- Can provide in-depth exploration of issues.
- Usually has face validity—the questions generally have a clear relationship to the outcomes being assessed.
- Can be combined with other techniques, such as surveys.
- The process allows faculty to uncover unanticipated results.

- Can provide insights into the reasons for participants' beliefs, attitudes, and experiences.
- Can be conducted within courses.
- Participants have the opportunity to react to each other's ideas, providing an opportunity to uncover the degree of consensus on ideas that emerge during the discussion.

Weaknesses of Focus Groups

- Generally provides indirect evidence about student learning.
- Requires a skilled, unbiased facilitator.
- Their validity depends on the quality of the questions.
- Results might not include the full array of opinions if only one focus group is conducted.
- What people say they do or know may be inconsistent with what they actually do or know.
- Recruiting and scheduling the groups can be difficult.
- Time-consuming to collect and analyze data.

46.3 Best Practices in Selecting Assessment Methods in Pharmacy Education

To select and apply the most valid and reliable assessment methods, pharmacy educators should take into consideration the following:

Select the appropriate assessment methods based on the literature and experience and make sure it is valid and reliable as well as approved by the school curriculum committee.

If you need to implement a new assessment method, take the approval from the department and curriculum committee.

Distribute assessment methods based on the course competencies and course learning outcomes as follows: select and apply the appropriate assessment methods for each course learning outcome; select and apply the appropriate assessment methods for each week's lectures, labs, and tutorials from week 1 to week 16 as an example.

Allocate marks for each assessment method (if applicable), for each week, and for each learning outcome based on the size of the learning outcome and other factors.

Revise the assessment methods on a yearly basis and modify them if needed.

46.4 Conclusion

This chapter has discussed indirect assessment methods in pharmacy education. This chapter includes background about indirect assessment methods. It also includes indirect assessment methods and their application in pharmacy education such as surveys and examples about them in pharmacy education, as well as the strengths and limitations of this type of assessment. Furthermore, it describes the best practices in selecting assessment methods in pharmacy education.

Reference

Allen, M.J., 2004. *Assessing academic programs in higher education* (p. 2013). Anker Publishing. Retrieved February, 6.

47

Assessment Methods in Pharmacy Education: Formative Assessment

47.1 Background

Formative assessment can be used to offer feedback for both teachers and learners to help the former decide how teaching should be adjusted and improved, and to help the latter learn better. Formative assessment has been defined as "activities undertaken by teachers—and by their students in assessing themselves—that provide information to be used as feedback to modify teaching and learning activities" (Dixson and Worrell, 2016). The purpose of formative assessment is to monitor student learning and provide ongoing feedback to staff and students. It is assessment for learning. If designed appropriately, it helps students identify their strengths and weaknesses, and can enable students to improve their self-regulatory skills so that they manage their education in a less haphazard fashion than is commonly found. It also provides information to the faculty about the areas students are struggling with so that sufficient support can be put in place. Formative assessment can be tutor led, peer or self-assessment. Formative assessments have low stakes and usually carry no grade, which in some instances may discourage the students from doing the task or fully engaging with it (Black and Wiliam, 2010; Garrison and Ehringhaus, 2007; Mavis, 2014; Alfadl, 2018).

A good formative assessment should be (Allen, 2004):

- Valid—directly reflects the learning outcome being assessed.
- Reliable—especially inter-rater reliability when subjective judgments are made.
- Actionable—results help faculty identify what students are learning well and what requires more attention.
- Efficient and cost-effective in time and money.
- Engaging to students and other respondents—so they'll demonstrate the extent of their learning.

DOI: 10.1201/9781003230458-52

- Interesting to faculty and other stakeholders—they care about results and are willing to act on them.
- Triangulation—multiple lines of evidence point to the same conclusion.

47.2 Key Components for Formative Assessment for Learning

DiVall et al. (2014) reports the following key components for formative assessment for learning:

Recognizing that formative assessment is a process, not a single activity, aimed at increased learning.

Moving focus away from achieving grades and onto the learning process, thereby increasing self-efficacy and reducing extrinsic motivation.

Conducting assessment activities during the teaching-learning process, not after, to assess achievement of predetermined outcomes.

Providing students with feedback to improve their learning and to focus on progress.

Providing faculty with information to guide instructional design and modify learning activities and experiences.

Allowing self-assessment to improve students' metacognitive awareness of their own learning.

47.3 Formative Assessment Methods and Their Application in Pharmacy Education

Literature reports the following examples of formative assessment methods (DiVall et al., 2014).

47.3.1 Prior Knowledge Assessment

A short quiz before or at the start of a class guides lecture content and informs students of weaknesses and strengths.

Limitations: Students may not be motivated to take the assessment seriously, requires flexible class time to respond.

Examples of Prior Knowledge Assessment in Pharmacy Education

Quiz about students' knowledge about pharmacovigilance and adverse drug reactions (ADRs) before the class on pharmacovigilance.

Quiz about students' knowledge about pharmacotherapy of asthma before the class on asthma.

47.3.2 Minute Paper

A writing exercise asking students what they thought was the most important information and what they did not understand.

It will take about a minute and is usually used at the end of class; it can be used at the end of any topic and in any course.

Can provide rapid feedback, and requires students to think and reason.

Limitations: Students may expect all items to be discussed, students may use it to get a faculty member to repeat information rather than introducing new information.

Examples of Minute Paper Assessment in Pharmacy Education

After the class on pharmacovigilance, the pharmacy educator may ask the students:

What have you learned today?

What didn't you understand today?

Do you have any questions?

47.3.3 Muddiest Point

Student response to a question regarding the most confusing point for a specific topic.

This quick and easy active learning activity asks students to identify the muddiest, or most confusing point in a lecture, class session, or assignment. By asking students to write this down and collecting their responses you can quickly identify the areas where your students are having difficulty. From there you can address those difficulties at the start of your next class. The effectiveness of this strategy hinges on addressing the muddiest points identified by the students. You can respond via email, your course management system, or with a short screencast or video.

Helps students acknowledge lack of understanding, identifies problem areas for the class.

Limitations: Emphasizes what students do not understand rather than what they do understand.

Examples of Muddiest Point Assessment in Pharmacy Education

After the beginning of pharmacovigilance:

> The pharmacy educator may give the students cards or white papers at the end of the class and ask them: What was the muddiest point in pharmacovigilance class? What was not clear or least clear to you?

Collect the students' responses and cards/papers.

Review and plan responses.

Post questions and answers online on the course page.

Answer questions at the beginning of next class.

Revise course contents if needed.

47.3.4 "Clickers" (Audience Response System)

A question asked at any time during a class to gauge learning.

Provides students/faculty with immediate feedback, and debriefing can improve the understanding of a concept.

Limitations: Uses up classroom time, and students may not be motivated to answer questions seriously.

Examples of "Clickers" (Audience Response System) Assessment in Pharmacy Education

During the class on pharmacovigilance, the pharmacy educator may ask the students to ask questions during class time. The pharmacy educator will give the students an immediate answer and feedback.

During the class on the pharmacotherapy of asthma, the pharmacy educator may ask the students to ask questions during class time.

The elements of effective feedback—the SMART system are:

Specific.

Measurable and Meaningful.

Accurate and Actionable.

Respectful.

Timely.

47.3.5 Case Studies (Problem Recognition)

Case analysis and response to case-related questions and/or identification of a problem.

Helps develop critical thinking and problem-solving skills, develops diagnostic skills.

Limitations: Time-consuming to create, takes considerable time for students to work on them.

Examples of Case Studies (Problem Recognition) in Pharmacy Education

Pharmacotherapy cases.

Clinical pharmacokinetics cases.

Others.

47.3.6 Formative Peer Assessment

Students are introduced to the assignment and criteria for assessment.

Students are trained and given practice on how to assess and provide feedback.

Students complete and submit a draft.

Students assess the drafts of other students and give feedback.

Students reflect on the feedback received and revise their work for final submission.

Assignments are graded by the instructor.

The instructor reflects on the activity with the class.

47.4 Assessment for Laboratory and Experiential Settings Strategies

A wide variety of formative assessment tools and methodologies can be used for in laboratory and experiential settings. One of the most commonly used assessment strategies in the laboratory setting is objective structured clinical examinations (OSCEs) (Miller, 1990; Epstein and Hundert, 2002; Beck et al., 1995).

47.5 Engage Students in Formative Assessment

- Explain the rationale behind formative assessment clearly—make it clear to students that through engaging with formative tasks they get to gain experience with their assessments, risk-free, and can develop far stronger skills in order to obtain better grades in the summative assessments.

- Create a link between summative and formative assessment—design formative assessments in such a way that they contribute to the summative task. This lowers the workload on the students and provides them with necessary feedback to improve their final performance. An example of such assessment is producing an essay plan, a structure of a literature review, part of the essay, or bibliography.

- Lower the number of summative assessments and increase the number of formative assessments—yet do not allow one single summative assessment to carry too much weight in the final grade.

47.6 Conclusion

This chapter has discussed formative assessment methods in pharmacy education. This chapter includes background about formative assessment methods. It also includes formative assessment methods and their application in pharmacy education such as prior knowledge assessment and examples about it in pharmacy education, as well as the strengths and limitations of this type of assessment. Furthermore, it describes the best practices in engaging students in formative assessment.

References

Alfadl, A.A., 2018. Assessment methods and tools for pharmacy education. In *Pharmacy education in the twenty first century and beyond* (pp. 147–168). Academic Press.

Allen, M.J., 2004. *Assessing academic programs in higher education* (p. 2013). Anker Publishing. Retrieved February, 6.

Beck, D.E., Boh, L.E. and O'Sullivan, P.S., 1995. Evaluating student performance in the experiential setting with confidence. *American Journal of Pharmaceutical Education*, 59, pp. 236–236.

Black, P. and Wiliam, D., 2010. Inside the black box: Raising standards through classroom assessment. *Phi Delta Kappan*, 92(1), pp. 81–90.

DiVall, M.V., Alston, G.L., Bird, E., Buring, S.M., Kelley, K.A., Murphy, N.L., Schlesselman, L.S., Stowe, C.D. and Szilagyi, J.E., 2014. A faculty toolkit for formative assessment in pharmacy education. *American Journal of Pharmaceutical Education*, 78(9).

Dixson, D.D. and Worrell, F.C., 2016. Formative and summative assessment in the classroom. *Theory into Practice*, 55(2), pp. 153–159.

Epstein, R.M. and Hundert, E.M., 2002. Defining and assessing professional competence. *JAMA*, 287(2), pp. 226–235.

Garrison, C. and Ehringhaus, M., 2007. *Formative and summative assessments in the classroom.* Available at: http://ccti.colfinder.org/sites/default/files/formative_and_summative_assessment_in_the_classroom.pdf

Mavis, B., 2014. Assessing student performance. In *An introduction to medical teaching* (pp. 209–241). Springer.

Miller, G.E., 1990. The assessment of clinical skills/competence/performance. *Academic Medicine,* 65(9), pp. S63–S67.

48

Objective Structured Clinical Examination (OSCE) in Pharmacy Education

48.1 Background

The objective structured clinical examination (OSCE) fulfils most of the previously mentioned criteria. It is a method of assessing a student's clinical competence that is objective rather than subjective, and in which the areas tested are carefully planned by the examiners. Objective structured clinical examination (OSCE) is an approach to the assessment of clinical competence in which the components of competence are assessed in a planned or structured way with attention being paid to the objectivity of the examination. (Harden and Gleeson, 1979). The examination consists of multiple, standard stations at which students must complete 1 to 2 specific clinical tasks, often in an interactive environment involving patient actors such as standardized patients (Harden and Gleeson, 1979; Sturpe, 2010).

48.2 Advantages of Objective Structured Clinical Examination (OSCE)

The main advantage of OSCE is its ability to assess candidates' multiple dimensions of clinical competencies (Sherazi, 2019):

- History taking
- Physical examination
- Medical knowledge
- Interpersonal skills
- Communication skills
- Professionalism
- Data gathering/information collection

DOI: 10.1201/9781003230458-53

- Understanding disease processes
- Evidence-based decision making
- Primary care management/clinical management skills
- Patient-centered care
- Health promotion
- Disease prevention
- Safe and effective practice of medicine

48.3 Steps of Objective Structured Clinical Examination (OSCE)

Kachur et al., 2012 suggested the following steps for Objective Structured Clinical Examination (OSCE):

Step 1. Identify Available Resources

a. Assemble a Team: One or more of the following:

Leader

Strong motivation to develop and implement the project.

Well connected to procure resources, including access to institutional or local clinical skills testing facilities.

Involved in medical school curriculum decision making.

Able to communicate well and create a team spirit.

Planner

Understands logistics of implementing OSCEs.

Is familiar with local conditions.

Can entertain multiple options for solving problems.

Administrator

Can implement OSCE-related tasks (e.g., scheduling, SP recruitment, photocopying of station materials, data entry).

Able to communicate well and create a team spirit.

Good at troubleshooting and problem solving.

Station Developer

Has relevant clinical experience.

Is familiar with performance standards.

Accepts editing.

Trainer

Understands SP and rater roles and case requirements.

Has teaching skills (e.g., provides constructive feedback) and can manage psychosocial impact of case portrayals.

Able to communicate well and create a team spirit.

Simulated Patients

At least one for each station.

Interested in taking on "educational" responsibilities.

Rater

At least one for each station.

Clear about OSCE goals and performance standards.

Committed to fair performance assessments.

Effective feedback provider.

Timer

At least one for each station.

Monitor

At least one for each station.

Data Manager

Data Analyst

Understands OSCE process.

Has psychometric skills.

Understands end-users of results (e.g., learners, program).

Program Evaluator

Understands OSCE process.

Is familiar with evaluation models (e.g., pre-/post-testing).

Can develop and analyze program evaluations (e.g., surveys, focus groups).

Best Practices: Assembling a Team

- Establish a clear common goal.
- Build a team with a variety of skills.
- Schedule regular meetings to build group identity.
- Create a common repository (i.e., shared drive, secure website) for meeting minutes, materials, and protocols.
- Look broadly for suitable sites and funding sources.
    b. Identify location.
    c. Identify sources of funding and support.

Step2. Agree on Goals and Timeline

Best Practices: OSCE Planning

Identify date and time of OSCE.

Make a timeline working backward from the OSCE date.

Start early to identify potential SPs and secure training times.

Identify potential location of OSCE early (clinic rooms, classrooms, or simulation center).

Secure participants' availability.

Step 3. Establish a Blueprint

Best Practices: Blueprinting

- Delineate core competencies.
- Establish performance criteria for each level of training.
- Ensure OSCE case patient age, gender, race, and prevalence of disease reflect actual clinical practice.
- Align OSCE skills and content assessed with current or new curricula.

Step 4. Develop Cases and Stations

Best Practices: Case Development

- Choose scenarios that are both common and challenging presentations for your learners.
- Ensure that cases represent the patient population in your clinical environment.
- Build specific goals and challenges into each scenario.
- Choose a post-encounter activity (i.e., feedback, supplementary exercise, or rest).
- Make sure it is possible to complete tasks in the time allotted.
- Organize a trial run with a variety of other learners to validate and fine tune cases.

Step 5. Create Rating Forms

Best Practices: OSCE Checklists

- Develop rating items based on the blueprint and ensure that a sufficient number of items are included to reliably assess competence within the targeted domains.
- Consider using both behavior-specific items and global rating items in OSCE rating forms to achieve a balance in terms of helping raters reflect important elements of their subjective responses and to enhance their objectivity in representing what happened during the encounter and providing learners with specific and more holistic feedback.
- Develop response options for behavior-specific items that reflect observable actions and strive to match the response options to the likely variation in performance of the learner population to maximize differentiation.

Step 6. Recruit and Train Simulated Patients (SPs)

Best Practices: SP Recruitment and Training

- Search for SPs through word-of-mouth strategies (e.g., by contacting other SPs, connecting with other SP trainers, talking to clinicians and acting teachers).
- Cast the right person for each case (i.e., physical appearance, psychological profile, availability, no contraindications).
- For high-stakes programs recruit and train alternates who can step in if needed (alternates can be cross-trained to provide coverage for multiple cases).

- Put SPs into learner's positions through role-play to enhance their understanding of the case (e.g., interactive and emotional impact of SP actions) and to promote an empathic approach to learners.
- Practice all aspects of the encounter (e.g., physical exam, feedback); do not leave SP performance to chance.
- Explore the psychological and physiological impact a case has on the SP to avoid toxic side effects (e.g., getting depressed from repeatedly portraying a depressed patient, getting muscle spasms from portraying a patient who has difficulty walking).
- Train all SPs who are portraying the same case (simultaneously or consecutively) at the same time to enhance consistency in case portrayal across SPs.

Step 7. Recruit and Train Evaluators

Best Practices: Evaluator Recruitment and Training for Rating and Feedback Tasks

- Select evaluators who are willing to adopt the program values, who are consistent in their ratings, and who don't have an ax to grind.
- Bring multiple evaluators together to jointly observe a learner's performance on tape or live, compare ratings, and discuss similarities and discrepancies. Practice giving feedback (if this is expected).
- Make raters aware of potential biases and rating mistakes.
- Provide written guidelines for rating items, evaluation scheme, and station objectives/teaching points.
- Post-OSCE, give feedback to raters about how their ratings compare with those of others (e.g., more or less lenient, lack of range).

Step 8. Implement the OSCE: Manage the Session

Step 9. Manage, Analyze, and Report Data

Step 10. Develop a Case Library and Institutionalize OSCEs

48.4 Applications of Objective Structured Clinical Examination (OSCE) in Pharmacy Education

Objective structured clinical examination (OSCE) can be applied and implemented in many ways such as the following:

Exit exam and comprehensive exam: to assess the readiness of students for pharmacy practice; this will also help them in the licensing exam.

Use a mini OSCE to assess the pharmacy and clinical skills of students in many courses such as: pharmaceutical care, pharmacy practice, pharmacotherapy, and others.

In the Introductory Pharmacy Practice Experiences (IPPEs) and Advanced Pharmacy Practice Experiences (APPEs).

48.5 Online Objective Structured Clinical Examination (OSCE)

Objective Structured Clinical Examination (OSCE) can be implemented online with the help of new technologies and training. Mock online OSCE is very important for pharmacy educators and students.

48.6 Challenges of Objective Structured Clinical Examination (OSCE)

Lack of resources.

Lack of funds.

Workforce issues.

Knowledge of pharmacy educators about OSCE.

Attitude of pharmacy educators towards OSCE.

Attitude of pharmacy students towards OSCE.

Resistance of pharmacy educators towards OSCE.

Resistance of pharmacy students towards OSCE.

University/school culture towards the implementation of OSCE.

Lack of training about OSCE.

Lack of motivation to train about OSCE.

Lack of technologies to implement online OSCE.

48.7 Conclusion

This chapter has discussed the Objective Structured Clinical Examination (OSCE). This chapter includes background about the OSCE. It also includes the advantages of OSCE. Moreover, the steps of OSCE and the best practices in each step are covered. Furthermore, it describes the applications of and barriers to implementing OSCE in pharmacy education.

References

Harden, R.M. and Gleeson, F.A., 1979. Assessment of clinical competence using an objective structured clinical examination (OSCE). *Medical Education*, 13(1), pp. 39–54.

Kachur, E.K., Zabar, S., Hanley, K., Kalet, A., Bruno, J.H. and Gillespie, C.C., 2012. Organizing OSCEs (and other SP exercises) in ten steps. In *Objective structured clinical examinations* (pp. 7–34). Springer.

Sherazi, M.H., 2019. Objective structured clinical examination introduction. In *The objective structured clinical examination review* (pp. 1–12). Springer.

Sturpe, D.A., 2010. Objective structured clinical examinations in doctor of pharmacy programs in the United States. *American Journal of Pharmaceutical Education*, 74(8).

Index

A

accreditation
  achievements, 83
  around world, 81
  background, 83–84
  definition, 80
  history, 80–81
  history of medical education
      accreditation, 80
  history of pharmacy education
      accreditation, 81
  importance, 79–80
  international accreditation, 81–82
  online pharmacies, 101
  online pharmacy education, 83
  pharmacy education curriculum,
      21–22
  standards, 82
active teaching strategies
  advantages of online pharmacy
      education, 86
  audience response system/clickers,
      209
  blended learning strategy, 213,
      285–292
  case-based learning teaching
      strategy, 210, 211, 214, 231–240,
      255
  community services-based learning
      teaching strategy, 34, 62, 214
  CP scheme, 212
  deliberative discussion, 209
  demonstration, 211
  discussion-based learning, 209
  discussion/debate, 211
  educational games, 211
  flipped classroom teaching strategy,
      213
  formative quizzes, 212
  interactive module, 211
  interactive-spaced education, 210
  interactive vodcast, 211
  interactive web-based learning, 210
  interview, 211
  journal club teaching strategy, 34, 42,
      61, 214
  lab/studio, 212
  learning stations, 211
  lecture-based/interactive lecture-
      based learning, 32, 213,
      279–283
  oral presentation, 212
  panel, 211
  patient simulation, 210
  problem-based learning teaching
      strategy, 210, 212, 225–228
  process-oriented guided inquiry
      learning (POGIL)/discovery
      learning, 210
  project-based learning strategy, 213
  role play teaching strategy, 33, 41, 47,
      212, 213, 241, 242, 245, 246–247,
      346
  self-learning/self-directed learning,
      34, 57–62, 214
  simulation-based learning teaching
      strategy, 210, 212, 213, 241–248
  team-based activity, 212
  team-based learning teaching
      strategy, 210, 217–224
  technology-enhanced active
      learning, 212
  traditional laboratory/benchtop
      experiences, 210
  video-based learning teaching
      strategy, 33, 213
  vodcast + hyperlinks, 211
  vodcast + pause activities, 210
  worksheet/problem set, 212
advanced certificates/board certificates
    for pharmacists
  Ambulatory Care Pharmacy
      Certificate, 14
  Board Certified Medication Therapy
      Management Specialists
      (BCMTMS), 13
  Board Pharmacy Specialties (BPS), 13

Compounded Sterile Preparations
Pharmacy Certificate, 14
Oncology Pharmacy Board
Certificates, 14
Pharmacist Independent Prescribing
Practice Certificate, 13
Solid Organ Transplantation
Pharmacy Certificate, 14
Advanced Pharmacy Practice
Experiences (APPE)
achievements, 49
assessment/evaluation methods,
48–49
best practices, 46–48, 49
challenges/recommendations, 49
curriculum, 3, 21, 31
definition, 46
distribution throughout curriculum,
45
journal club teaching strategy, 48
objective structured clinical
examination, 347
project-based learning strategy, 48
role play teaching strategy, 47
simulation-based training, 47
simulation cases, 47
simulation consultation services, 47
simulation drug information
services, 48
simulation medication errors
reporting, 48
simulation pharmacovigilance
and adverse drug reactions
reporting, 48
simulation prescriptions/orders, 47
simulation/virtual learning, 47
virtual resources, 48
adverse drug reactions (ADRs), 14, 27,
48, 102, 115, 119, 120, 158
Ancient Babylonia, 93
Ancient China, 93
Ancient Egypt, 93
assessment/evaluation methods
Advanced Pharmacy Practice
Experiences (APPE), 48
diagnostic assessment, 35, 42, 308
direct assessment, 308, 317–324
formative assessment, 34–35, 42, 48,
333–339

indirect assessment, 308, 325–331
Introductory Pharmacy Practice
Experiences (IPPE), 48
key features, 309
objective structured clinical
examination, 341–348
online pharmacy education, 86, 89
performance measures, 311–312
portfolios, 312–313
practicals/labs/tutorials, 42
rationality of, 308–309
selection of tools, 313–314
standardized exams, 310–311
strengths/limitations, 310–313
summative assessment, 42, 48, 308
terminologies, 307
theory courses, 34–35

B

Blackboard, 52, 286
blended learning strategy
active teaching strategies, 213
barriers, 291
best practices for, 290–291
definition, 285
enriched virtual model, 289
flex model, 289
flipped classroom model, 288
history, 285–286
individual rotation model, 288
lab rotation model, 288
a la carte model, 289
models, 288–289
purpose, 288
station rotation model, 288
theory courses, 33
tips for implementation, 289

C

case-based learning teaching strategy
active teaching strategies, 211, 214
barriers, 233
definition, 231
forms of, 233–234
history, 231
importance, 231–232

long case example in
pharmacotherapy, 234–240
practicals/labs/tutorial, 47
purpose, 232
self-learning/self-directed learning,
61
theory courses, 34
tips for designing effective, 232
tips for implementation, 232–233
Cisco WebEx Teams:, 31, 38, 46, 52, 76,
105, 108, 115, 126, 130, 267
community services
awareness campaigns, 72
background, 69, 72
developing communication skills, 70
developing drug information/patient
counseling skills, 70
developing empathy, 70
developing new friendships, 70–71
drug/medications/poison
information services, 72
faculty mentors, 70
gaining cultural competence, 70
importance, 69–71
opportunities outside of pharmacy,
71
tools/ideas, 71–72
using social media, 71
using WhatsApp, 71
using YouTube, 71
community services-based learning
teaching strategy
active teaching strategies, 214
self-learning/self-directed learning,
62
theory courses, 34
competencies-based education, 23–24,
29
computer-assisted learning/computer-
based learning
applications, 301–303
barriers, 303–304
Blackboard, 302
computers, 302
definition, 299
history, 300
importance, 300–301
internet, 301–302
laptops, 302

Learning Management Systems, 302
Moodle, 302
purpose, 300
video conferencing platforms, 303
webinar, 303
computers, 51, 75, 88, 130, 137, 149, 155,
196, 199, 302
continuing professional development
(CPD)
Ambulatory Care Pharmacy
Certificates, 140
anticoagulation certificate, 138
barriers to effective, 136–137
benefits, 135
Board Certified Medication Therapy
Management Specialists, 139
Board Pharmacy Specialties, 139
certificates and board certificates,
138–140
Compounded Sterile Preparations
Pharmacy Certificates, 140
definition, 57, 62
diabetes management certificate,
138
free courses, 140
history, 133
importance, 141
life-long learning, 134
medication safety certificate, 138
mobile health, 114
Oncology Pharmacy Board
Certificates, 139
pain management certificate, 138
pharmacist independent prescribing
practice certificate, 139
pharmacogenomics certificate, 138
principles, 135
rationality of, 134–136
research skills/competencies, 197
Solid Organ Transplantation
Pharmacy Certificates, 140
steps, 136
technologies/tools, 137
continuous pharmacy education
applying learning to teaching/
practice, 67
applying self-learning process
approach, 67
attending academic advising, 66

attending online patient care services
  programs, 66
background, 63, 67–68
best practices, 66–67
course design/online delivery, 65
designing/developing goals of, 67
developing better organization and
  planning skills, 64
documentation, 67
identifying activities, courses,
  programs, workshops,
  conferences, and other
  activities, 67
importance, 63–66
improve/update knowledge skills, 64
improving communication skills, 64
improving technology-related
  knowledge/skills, 64
leadership skills development, 64
learning new assessment and
  evaluation methods, 64
learning new teaching strategies, 63
maintaining electronic portfolio, 66
needs identification, 66
reflection, 67
sharing with colleagues, 67
using electronic files, 66
using Learning Management
  Systems, 65
using social media, 65–66
using video conferencing platforms,
  65
using Webinar, 65
using YouTube, 65

D

diagnostic assessment
  Advanced Pharmacy Practice
    Experiences (APPE), 48
  definition, 35, 42, 308
  Introductory Pharmacy Practice
    Experiences (IPPE), 48
  practicals/labs/tutorials, 42
  theory courses, 34–35
digital/technology literacy, 77
direct assessment
  application in pharmacy education,
    317–322

background, 317
best practices, 323
definition, 308
embedded assignments/course
  activities, 320–321
locally developed tests, 319–320
portfolios, 321–322
published tests/standardized tests,
  318–319
distance education
  definition, 5, 274
  history, 6
  pharmacy education and, 6
  phases of development, 6
drug information/counseling services,
  95
drug related problems (DRPs), 119, 126

E

educational games
  active teaching strategies, 211
  definition, 265
  history, 265–266
  importance, 266
  in pharmacy education, 267–269
  purpose, 267
e-learning, 274

F

Facebook, 53, 105, 106
feedback, 5, 20, 32, 33, 35, 38, 42, 47, 49,
  60, 213, 220, 221, 222, 232, 244,
  246, 247, 250, 251, 253, 255, 261,
  262, 265, 270, 274, 275, 277, 280,
  282, 289, 291, 296, 299, 307, 309,
  311, 312, 328, 333, 334, 335, 336,
  337, 338, 343, 345, 346
flipped classroom teaching strategy
  active teaching strategies, 213
  barriers, 263
  blended learning strategy, 288
  case reports, 261
  definition, 255
  dry lab, 261
  history, 255–256
  homework, 260–261
  importance, 256

pre-classroom activities, 259
preparation, 259
principles, 258
purpose, 257
readiness assurance, 260
reasons for using, 257
review session, 261
self-learning/self-directed learning,
    62
structure, 258–262
student progress assessment, 261
theory courses, 34
tips for, 258
tips for implementation, 262–263
worksheet, 260–261
formative assessment
    Advanced Pharmacy Practice
        Experiences (APPE), 48
    application in pharmacy education,
        334–337
    audience response system/clickers,
        336
    background, 333–334
    case studies/problem recognition,
        336–337
    definition, 307
    formative peer assessment, 337
    Introductory Pharmacy Practice
        Experiences (IPPE), 48
    key components, 334
    laboratory and experiential settings,
        337
    minute paper, 335
    muddiest point, 335–336
    practicals/labs/tutorials, 42
    prior knowledge assessment, 334–335
    student engagement, 337–338
    theory courses, 34–35

G

Google Meet, 31, 38, 46, 52, 76, 105, 108,
    115, 126, 130, 267

I

indirect assessment
    application in pharmacy education,
        326–330

background, 325
best practices, 330
definition, 308
focus groups, 329–330
interviews, 327–329
surveys, 326–327
Instagram, 53, 105, 107
internet, 51, 74–75, 87–88, 96, 99, 129, 137,
    149, 155, 196, 199, 301–302
Introductory Pharmacy Practice
        Experiences (IPPE)
    achievements, 49
    assessment/evaluation methods,
        48
    best practices, 46–48, 49
    challenges/recommendations, 49
    curriculum, 3, 21, 31
    definition, 45
    distribution throughout curriculum,
        45
    journal club teaching strategy, 48
    objective structured clinical
        examination, 347
    project-based learning teaching
        strategy, 48
    role play teaching strategy, 47
    simulation-based training, 47
    simulation cases, 47
    simulation consultation services,
        47
    simulation drug information
        services, 48
    simulation medication errors
        reporting, 48
    simulation pharmacovigilance
        and adverse drug reactions
        reporting, 48
    simulation prescriptions/orders, 47
    simulation/virtual learning, 47
    virtual resources, 48

J

journal club teaching strategy
    active teaching strategies, 214
    practicals/labs/tutorial, 42
    self-learning/self-directed learning,
        61
    theory courses, 34

L

laptops, 51, 75, 88, 130, 137, 149, 196, 199, 302
learning environment, 77
Learning Management Systems (LMS), 52, 65, 75–76, 286, 302
learning styles
    4MATlearning style, 207
    accommodating learning style, 205
    assimilating learning style, 205
    aural/auditory learning style, 204
    converging learning style, 205
    definition, 203
    diverging learning style, 206
    frameworks, 203–207
    Grasha-Reichmann Student Learning Style Scale, 206–207
    Gregorc Style Delineator, 206
    Honey and Mumford learning style questionnaire, 206
    kinesthetic learning style, 204
    Kolb's experiential learning style theory, 205–206
    Pharmacist Inventory of learning styles, 206
    read/write learning style, 204
    VARK learning style, 203–204
    visual learning style, 203
lecture-based/interactive lecture-based learning
    active teaching strategies, 213
    advantages, 280
    barriers, 282–283
    definition, 279
    disadvantages, 280–281
    history, 279
    purpose, 281
    theory courses, 32–33
    tips for designing effective, 281–282
    tips for implementation, 282
life-long learning, 25, 67, 79, 134

M

mail-order pharmacy, 95, 99–100
managed learning environment (MLE), 273
massive open online courses (MOOCs)
    challenges, 296
    cMOOC model, 295
    definition, 293–294
    history, 294
    importance, 294–295
    models, 295
    in pharmacy education, 295
    purpose, 295
    xMOOC model, 295, 296
medication errors, 120–121
medication safety
    adverse drug reactions, 14, 27, 48, 115, 119, 120
    background, 119
    certificate, 138
    competencies, 27
    medication errors, 120–121
    medication misuse/abuse, 121
    mobile health, 115
    online pharmacies, 158
    pharmacovigilance, 119
    prescribing, 148
    self-medications, 121
    substandard/counterfeit medications, 122
medication therapy management (MTM), 125, 129
Microsoft Teams, 31, 38, 46, 52, 76, 105, 108, 115, 130, 267
mobile health (m-Health/mhealth)
    apps not regulated, 117
    apps subject to enforcement discretion, 116
    clinical calculators, 113–114
    continuing education and professional development, 114
    diagnostic support tools/point-of-care diagnostics, 114
    drug information resources, 113
    guidelines, 114
    history, 111–112
    impact in pharmacy practice, 113–115
    literature databases, 114
    medical uses, 112
    medication safety, 115
    medicines availability, 115
    quality standards, 117
    reasons for using, 113
    regulated apps, 116

regulations, 116–117
social media, 115
terminologies, 111
Moodle, 2, 52, 286

N

net books, 51–52, 75, 88, 130, 137, 149, 155

O

objective structured clinical
    examination (OSCE)
    advantages, 341–342
    application in pharmacy education,
        346–347
    background, 341
    best practices, 344–346
    blueprinting, 344
    case development, 345
    case library development, 346
    challenges, 347
    checklists, 345
    data analysis, 346
    data management, 346
    data reports, 346
    evaluator recruitment and training,
        346
    implementation, 346
    institutionalization, 346
    online implementation, 347
    planning, 344
    simulated patient recruitment/
        training, 345–346
    steps, 342–346
    team, 342–344
online/digital library, 54, 65, 76, 88, 303
online learning, 38, 46, 77, 85, 89,
        273–274, 285, 288, 289, 293
online patient care
    background, 125
    barriers to effective, 130–131
    dispensing medications, 125–126
    drug related problems, 126
    management plan, 126
    medication review, 125, 127–129
    medication therapy management, 129
    patient assessment, 126
    regulations, 129

technologies/tools, 129–130
online pharmacies
    access to health care services, 153
    advantages, 153–154
    benefits, 154
    characteristics of, 100–101
    communication skills, 155
    consultative services, 102
    convenience, 154
    cost savings, 153
    definition, 99
    disadvantages, 103
    disadvantages/problems, 154–155
    environmental benefits, 154
    history, 96, 99
    importance, 101
    infrastructure, 155
    laws/regulations, 101
    licensure/policy maintenance, 102
    mail-order pharmacy, 95, 99–100
    medication safety practice, 158
    nonprescription medications/self-
        medication practice, 156–157
    over-the-counter products, 102
    patient information, 102
    prescribed medications, 156
    prescriptions, 102
    quality improvement programs, 103
    quality of online dispensing, 156–157
    quality of online medications safety
        practice, 158
    quality of online prescribing, 157
    quality standards, 101–103
    safety, 153
    storage/shipment, 102
    technical support, 155
    technologies/tools, 155
    technology skills, 154, 155
    training, 155
    workload, 155
online pharmacy education
    accessible/affordable, 86
    accreditation, 83
    advantages, 85–87
    communication and collaboration, 87
    community services, 87
    competencies, 88–89
    continuity of teaching and learning,
        85

convenience, 85
cost, 85
degrees, 10
disadvantages, 87–89
distance education and, 6
distraction, 89
environmental benefits, 87
history, 6
importance, 7
improving self-learning skills, 86
improving technology knowledge, 85
improving technology skills, 86
infrastructure, 86, 87
lack of self-motivation, 89
negative attitude, 89
quality standards, 89
safety, 86
schedules, 86
student engagement, 87
teaching strategies, 86
technologies/tools, 51–55, 87–88
time management, 87
workforce issues, 87
workload, 89
online pharmacy education certificate
        courses/programs
    advanced certificates and board
        certificates for pharmacists,
        13–14
    anticoagulation certificate, 12
    diabetes management certificate, 12
    medication safety certificate, 12, 13
    pain management certificate, 12
    pharmacogenomics certificate, 12
    pharmacy education certificate
        courses and programs, 12

P

Papyrus Ebers, 93
patient information, 102
peer reviews, 191–193
pharmacovigilance, 48, 119
pharmacy education
    access/equitable access, 73–78
    barriers, 270–271
    cost, 73–74
    definition, 3
    distance education and, 6

educational games, 267–269
history, 4–5
location of pharmacy schools/
        departments, 74
massive open online courses,
        295–296
number of pharmacy schools/
        departments, 74
research, 169
role of internet, 74–76
technologies/tools, 75–76
tips for designing effective, 269–270
tips for implementation, 270
pharmacy education competencies
    achievements, 28
    challenges/recommendations,
        28–29
    cognitive skills, 25, 79
    collaboration, 25
    communication skills, 25, 79, 88
    community services activities, 28, 79
    competencies-based education,
        23–24, 29
    development of, 23
    education skills, 25
    ethical principles, 27–28, 79
    health promotion, 28, 79
    knowledge skills, 24–25, 79
    lack of/insufficient training, 89
    leadership abilities/skills, 26
    leadership skills development, 79
    learning outcomes and, 24–28
    legal requirements, 27–28
    life-long learning, 25, 79
    management skills, 26
    medication safety, 27, 79
    patient care services, 26–27
    personal/professional development,
        25
    pharmaceutical industries, 26, 79
    pharmacist care services, 26–27, 79
    prescribing practice skills, 27, 79
    professional responsibilities, 27–28
    research skills, 28, 80, 195, 197
    technology skills, 28, 80, 89
pharmacy education curriculum
    accreditation, 21–22
    achievements, 20
    administrative courses, 19

Advanced Pharmacy Practice
      Experiences (APPE), 3, 21, 31
bacteriology, 18
beginning, 18
challenges/recommendations,
      20–22
chemistry, 18
clinical pharmacy, 19
definition, 17
distribution of Practicals/
      Laboratories and tutorials
      throughout curriculum, 37
economic courses, 19
extracurricular activities, 21
Introductory Pharmacy Practice
      Experiences (IPPE), 3, 21, 31
laboratory training, 18
online courses, 19
online curriculum reform related
      issues, 19–20
online teaching/assessment, 21
pharmaceutical sciences, 18
pharmacognosy, 18
pharmacology, 18
phases of development, 17–19
physiological chemistry, 18
practical/laboratory courses, 20
quality standards, 21–22
social courses, 19
technologies facilities, 21
theory courses, 31–36
pharmacy education degrees
Bachelor of Pharmacy/
      Pharmaceutical Sciences, 9
Diploma in Pharmacy, 9
Doctorate/PhD Level, 11
Doctor of Pharmacy (PharmD), 9, 11
master level, 11
Master's in Pharmacy, 10
Postgraduate Diploma, 9
postgraduate diploma, 11
pharmacy education programs
certificate courses, 12
postgraduate programs, 11
undergraduate programs, 10
pharmacy education teaching strategies
active teaching strategies, 33, 34, 41,
      47, 61, 62, 86, 209–215, 241, 242,
      245, 246–247, 346

blended learning strategy, 213,
      285–292
case-based learning teaching
      strategy, 210, 214, 231–240, 255
community services-based learning
      teaching strategy, 34, 62, 214
computer-assisted learning/
      computer-based learning,
      299–304
educational games, 211, 265–272
flipped classroom teaching strategy,
      34, 62, 213, 255–264, 288
journal club teaching strategy, 34, 42,
      61, 214
lecture-based/interactive lecture-
      based learning, 213, 279–283
massive open online courses,
      293–297
practicals/labs/tutorials, 39–42
problem-based learning teaching
      strategy, 210, 212, 225–228
project-based learning teaching
      strategy, 213–214, 249–254, 255
role play teaching strategy, 33, 41, 47,
      213, 241, 242, 245, 246–247, 346
self-learning/self-directed learning,
      34, 57–62, 214
simulation-based learning teaching
      strategy, 210, 212, 213, 241–248
team-based learning teaching
      strategy, 210, 217–224
video-based learning teaching
      strategy, 213
web-based learning, 273–278
pharmacy practice
Ancient Babylonia, 93
Ancient China, 93
Ancient Egypt, 93
current practices, 94
drug information/counseling
      services, 95
early Greek philosopher, 94
first apothecary shops, 94
history, 93–94
history of distance/remote and
      online, 95–96
importance, 96
mail-order pharmacy, 95
online pharmacies, 96

pharmacopoeia, 94
separation of pharmacy and
    medicine, 94
telepharmacy, 95–96
Pinterest, 105
practicals/labs/tutorials
    access/equitable access, 77
    achievements, 43
    assessment/evaluation methods, 42
    best practices, 38
    challenges/recommendations, 43
    distribution throughout curriculum,
        37
    journal club teaching strategy, 41
    practicals/labs recording, 39, 40, 41
    project-based learning teaching
        strategy, 39, 40
    role play teaching strategy, 41
    simulation/virtual learning, 41
    teaching strategies for biomedical
        sciences related, 39–40
    teaching strategies for
        pharmaceutical sciences
        related, 40
    teaching strategies for social/
        administrative/behavioral and
        clinical sciences, 41–42
    virtual laboratories, 39, 40
    virtual resources, 39–40, 41
prescribing
    barriers to effective online
        prescribing, 149–150
    communication skills indicators,
        146–147
    competencies, 27, 79, 148–149
    definition, 143
    dependent prescribing/collaborative
        prescribing, 144
    diagnosing, 143
    diagnosis indicators, 147
    gathering information indicators, 147
    history, 144–145
    independent prescribing, 143–144
    management indicators, 147–148
    monitoring parameters, 148
    nonmedical prescribing, 143
    patient education/counseling, 148
    practice skills, 27, 79
    quality standards, 146–147

rationality of, 145–146
regulations, 150
supplementary prescribing, 144
technologies/tools, 149
terminologies, 143–144
problem-based learning teaching
    strategy
    active teaching strategies, 210, 212
    barriers, 229
    definition, 225
    disadvantages, 228
    history, 225–226
    importance, 226
    process, 227
    purpose, 226–227
    self-learning/self-directed learning,
        61
    theory courses, 33
    tips for designing effective, 228
    tips for implementation, 229
project-based learning teaching strategy
    active teaching strategies, 213
    Advanced Pharmacy Practice
        Experiences (APPE), 48
    barriers, 254
    definition, 249
    elements, 250–251
    history, 249
    importance, 249–250
    Introductory Pharmacy Practice
        Experiences (IPPE), 48
    practicals/labs/tutorial, 39, 40, 41
    purpose, 250
    self-learning/self-directed learning,
        61
    support from educators, 252
    theory courses, 33, 34
    tips for designing effective, 253
    tips for implementation, 253

R

research
    anonymity, 166
    bias, 167
    clinical trials, 163
    controlled trials, 164
    convenience, 166
    cost, 166

data accuracy, 166
data analysis, 166
Declaration of Helsinki, 164–165
disadvantages, 167
equity access, 167
flexibility, 166
follow-up, 167
history of medical research, 163–165
history of online research, 165
importance of online research,
165–166
improved access to populations, 166
insulin discovery, 164
limited questions, 167
number of participants, 167
Nuremburg Code, 164
penicillin discovery, 164
pharmacy education research, 169
research skills, 28, 80
sample sizes, 166
sampling techniques, 167
smallpox vaccine, 163
technical problems, 167
terminologies, 169–175
time saving, 166
vaccine discovery, 164
research barriers
absence of professional ethical
committees, 198
inadequate research skills/
competencies, 197
lack/absence of technologies/tools,
199
lack of collaboration, 198
lack of funding, 197
lack of infrastructure, 197
lack of interest/motivation, 197–198
lack of research facilities, 198
lack of time, 198
negative attitudes, 198
poor communication, 198
research facilitators
attitude, 195
collaboration, 196
communication, 196
funding, 195
Interest, 195
research skills/competencies, 195
teamwork, 196

technologies/tools, 196–197
time, 195
research implementation
approval, 184
background, 183
barriers, 188
data analysis procedure/statistical
analysis, 187
expected outcomes, 184
justification, 183
methodology, 184–185
objective/questions, 184
references style, 187
research area, 183
sampling procedure, 186
significance, 184
study conclusion, 187
study discussion, 187
study hypothesis, 184
study results, 187
study tool, 186
tips for implementation, 183–187
tips for publishing research, 187
topic background, 184
topic selection, 183, 186–187
research methods/methodology
analytic studies, 178, 185–186
classification based on data collection
type, 178
classification based on data sources,
177
classification based on outcome
exposure, 178
classification based on research
purpose, 178
classification of analytic studies, 179
classification of pharmacoeconomic
studies, 179–180
cost-benefit analysis, 179
cost-effectiveness analysis, 180
cost-minimization analysis, 179
cost of illness analysis, 180
cost utility analysis, 180
descriptive studies, 178, 185
experimental (intervention studies),
179, 186
importance, 177
issues, 177–180
mixed method studies, 178, 185

non-experimental (observational
    studies), 179, 185
primary research, 177
prospective study design, 178, 185
qualitative studies, 178, 185
quantitative studies, 178, 185
quasi-experimental, 179, 186
research implementation, 184–185
retrospective study design, 178, 185
review, 185
secondary research, 177
simulation studies, 178, 185
study methods/methodology for
    online research, 180
research quality
    animal pre-clinical studies, 193
    diagnostic accuracy studies, 193
    economic evaluation studies, 193
    expected outcomes, 190
    importance, 189
    indicators/criteria, 189–191
    justification, 190
    methodology, 190
    multivariate prediction models, 193
    objective/questions, 190
    observational studies in
        epidemiology, 193
    peer reviews, 191
    quality improvement studies, 193
    randomized controlled trials (RCTs),
        193
    study discussion, 191
    study hypothesis, 190
    study results, 191
    study topic, 189
    systematic reviews and meta-
        analyses, 193
    tools for assessing, 193–194
    topic background, 190
    web-based surveys, 194
research terminologies
    bias, 171
    case control study, 169–170
    case reports, 170
    case series, 170
    clinical trials, 170
    cohort study, 170
    cost-benefit analysis, 172
    cost-effectiveness analysis, 172

cost-minimization analysis, 172
cost of illness analysis, 173
cost utility analysis, 173
critical review, 173
cross-sectional studies, 169
descriptive studies, 169
ecological studies, 170–171
literature review, 173
meta-analysis, 174
mixed methods review, 174
pharmacoeconomics research, 172
pharmacy practice research, 169
prospective studies, 171
qualitative studies, 171
quantitative research, 171
rapid review, 173
retrospective studies, 171
sample sizes, 171
scoping review, 173
systematic review, 174
systematic search/review, 174
traditional (narrative) literature
    review, 173
umbrella review, 174
role play teaching strategy, 33, 41, 47,
    212, 213, 241, 245, 246–247, 346
    role play teaching strategy, 212, 242

S

sample method, 186
sample sizes, 166, 167, 171, 186, 190, 192
sampling procedure, 186
self-learning/self-directed learning
    active teaching strategies, 214
    best practices, 58–61
    case studies discussion teaching
        strategy, 61
    community services-based learning
        teaching strategy, 61
    definition, 57
    flipped classroom teaching strategy,
        61
    journal club teaching strategy, 61
    problem-based learning teaching
        strategy, 61
    project-based learning teaching
        strategy, 61
    rationality of, 57–58

responsibilities and roles of
pharmacy educators, 60
responsibilities of pharmacy schools/
departments, 60
responsibilities of pharmacy
students, 60–61
teaching strategies, 34, 61–62
team-based learning teaching
strategy, 61
theory courses, 34
seminars, 34, 214
Short Hand, 5
simulation-based learning teaching
strategy
active teaching strategies, 212, 213
Advanced Pharmacy Practice
Experiences (APPE), 47–48
applications, 245–246
barriers, 247
definition, 241
history, 242
importance, 242–244
Introductory Pharmacy Practice
Experiences (IPPE), 47–48
practicals/labs/tutorial, 41
purpose, 244
theory courses, 33
tips for designing effective, 246
tips for implementation, 246–247
types, 244
smartphones, 51–52, 75, 88, 111, 130, 137,
149, 196, 199
Snapchat, 105
social media
Cisco WebEx Teams:, 31, 38, 46, 52, 76,
105, 108, 115, 126, 130, 267
definition, 111
Facebook, 53, 105, 106
Google Meet, 31, 38, 46, 52, 76, 105,
108, 115, 126, 130, 267
guidelines for use by pharmacists,
108–109
history, 105
impact in pharmacy practice, 106–108
importance, 109
Instagram, 53, 105, 107
Microsoft Teams, 31, 38, 46, 52, 76,
105, 108, 115, 130, 267
mobile health, 115

online patient care, 130
Pinterest, 105
risks, 108
Snapchat, 105
Twitter, 53, 105, 106–107
use in continuous pharmacy
education, 65
using, 106
using in community services, 71
video conferencing platforms, 52–53,
65, 76, 107–108, 137, 149, 196, 199,
286, 303
webinar, 52–53, 65, 76, 107–108, 137,
149, 196, 199, 286, 303
WhatsApp, 53, 105, 107, 122, 137, 149
Wikipedia, 105
YouTube, 65, 105, 107
Zoom, 31, 38, 40, 46, 52, 76, 108, 115,
126, 130, 267
summative assessment
Advanced Pharmacy Practice
Experiences (APPE), 48
definition, 308
Introductory Pharmacy Practice
Experiences (IPPE), 48
practicals/labs/tutorials, 42
theory courses, 34–35
support, 78, 89

T

tablets, 51–52, 75, 88, 130, 137, 149, 155,
196, 199
team-based learning teaching strategy
accountability, 219
active teaching strategies, 210
barriers, 223–224
definition, 217
elements, 219
feedback, 220
group management, 219
history, 217–218
importance, 218
learning and team development, 220
principles, 219–220
purpose, 218
self-learning/self-directed learning,
61
theory courses, 33

tips for designing course, 221–222
tips for implementation, 220
tips for implementation online,
   222–223
technologies/tools, *see also* social media
   achievements, 54
   Blackboard, 52, 286, 302
   challenges/recommendations, 54
   computers, 51, 75, 88, 130, 137, 149,
      155, 196, 199, 302
   Facebook, 53, 105, 106
   Instagram, 53, 105, 107
   internet, 51, 74–75, 87–88, 96, 99, 129,
      137, 149, 155, 196, 199, 301–302
   laptops, 51, 75, 88, 130, 137, 149, 196,
      199, 302
   Learning Management Systems, 52,
      65, 75–76, 286
   Moodle, 2, 52, 286, 302
   net books, 51–52, 75, 88, 111, 130, 137,
      149, 155
   online/digital library, 54, 65, 76, 88,
      303
   in online pharmacy education, 51–55
   smartphones, 51–52, 75, 88, 111, 130,
      137, 149, 196, 199
   tablets, 51–52, 75, 88, 111, 130, 137, 149,
      155, 196, 199
   Twitter, 53, 105, 106–107
   video conferencing platforms, 52–53,
      65, 76, 107, 130, 137, 149, 196, 199,
      286, 303
   wearable technologies, 53, 76–77, 130,
      137, 149, 197, 199
   webinar, 52–53, 65, 76, 107, 130, 137,
      149, 196, 199, 286, 303
   WhatsApp, 53, 71, 105, 107, 122, 130,
      137, 149, 197, 199
   YouTube, 53, 65, 105, 107
telepharmacy, 95–96
theory courses
   achievements, 35
   assessment/evaluation methods,
      34–35
   best practices, 31–32
   blended learning strategy, 33
   case studies discussion teaching
      strategy, 34

challenges/recommendations,
   35–36
community services-based learning
   teaching strategy, 34
flipped classroom strategy, 34
journal club teaching strategy, 34
lecture-based/interactive lecture-
   based learning, 32–33
problem-based learning teaching
   strategy, 33
project-based learning teaching
   strategy, 34
role play teaching strategy, 33
self-directed learning, 34
seminars, 34
simulations methods teaching
   strategy, 33
teaching strategies, 32
teaching strategy, 34
team-based learning teaching
   strategy, 33
video-based learning teaching
   strategy, 33
traditional teaching strategies, 209
Twitter, 53, 105, 106–107

V

video-based learning teaching
   strategy
   active teaching strategies, 213
   theory courses, 33
video conferencing platforms, 52–53, 65,
   76, 107, 130, 137, 149, 196, 199,
   286, 303
virtual learning environment (VLE),
   273

W

wearable technologies, 53, 76–77, 130,
   137, 149, 197, 199
web-based learning
   advantages, 275–276
   barriers, 277–278
   disadvantages, 276
   history, 274–275
   purpose, 276, 277

terminologies, 273–274
tips for designing effective, 277
webinar, 52–53, 65, 76, 107–108, 137, 149,
    196, 199, 286, 303
WhatsApp, 53, 71, 105, 107, 122, 130, 137,
    149, 197, 199
Wikipedia, 105
workforce issues, 77–78, 87

Y

YouTube, 53, 65, 105, 107

Z

Zoom, 31, 38, 40, 46, 52, 76, 105, 108, 115,
    126, 130, 267

Printed in the United States
by Baker & Taylor Publisher Services